MONOGRAPHS ON THE
PHYSICS AND CHEMISTRY OF
MATERIALS

General Editors
R. J. BROOK, ANTHONY CHEETHAM,
ARTHUR HEUER, SIR PETER HIRSCH,
TOBIN J. MARKS, JOHN SILCOX,
D. G. PETTIFOR, MANFRED RUHLE,
MATTHEW V. TIRRELL, VACLAV VITEK

MONOGRAPHS ON THE PHYSICS AND CHEMISTRY OF MATERIALS

Quantum theory of collective phenomena G. L. Sewell
Experimental high-resolution electron microscopy (Second edition)
   J. C. H. Spence
Experimental techniques in low-temperature physics Guy K. White
Principles of dielectrics B. K. P. Scaife
Surface analytical techniques J. C. Rivière
Basic theory of surface states Sydney G. Davison and Maria Steślicka
Acoustic microscopy Andrew Briggs
Light scattering: principles and development W. Brown
Quasicrystals: a primer (Second edition) C. Janot
Interfaces in crystalline materials A. P. Sutton and R. W. Balluffi
Atom probe field ion microscopy M. K. Miller *et al.*
Rare-earth iron permanent magnets J. M. D. Coey (ed.)
Statistical physics of fracture and breakdown in disordered systems
   B. K. Chakrabarti and L. G. Benguigui

Statistical Physics of Fracture and Breakdown in Disordered Systems

BIKAS K. CHAKRABARTI
Saha Institute of Nuclear Physics, Calcutta

L. GILLES BENGUIGUI
Technion-Israel Institute of Technology, Haifa

CLARENDON PRESS · OXFORD
1997

This book has been printed digitally and produced in a standard specification in order to ensure its continuing availability

OXFORD
UNIVERSITY PRESS

Great Clarendon Street, Oxford OX2 6DP

Oxford University Press is a department of the University of Oxford.
It furthers the University's objective of excellence in research, scholarship,
and education by publishing worldwide in

Oxford New York

Auckland Cape Town Dar es Salaam Hong Kong Karachi
Kuala Lumpur Madrid Melbourne Mexico City Nairobi
New Delhi Shanghai Taipei Toronto
With offices in
Argentina Austria Brazil Chile Czech Republic France Greece
Guatemala Hungary Italy Japan South Korea Poland Portugal
Singapore Switzerland Thailand Turkey Ukraine Vietnam

Oxford is a registered trade mark of Oxford University Press
in the UK and in certain other countries

Published in the United States
by Oxford University Press Inc., New York

© B. K. Chakrabarti and L. G. Benguigui 1997

The moral rights of the author have been asserted

Database right Oxford University Press (maker)

Reprinted 2009

All rights reserved. No part of this publication may be reproduced,
stored in a retrieval system, or transmitted, in any form or by any means,
without the prior permission in writing of Oxford University Press,
or as expressly permitted by law, or under terms agreed with the appropriate
reprographics rights organization. Enquiries concerning reproduction
outside the scope of the above should be sent to the Rights Department,
Oxford University Press, at the address above

You must not circulate this book in any other binding or cover
And you must impose this same condition on any acquirer

ISBN 978-0-19-852056-6

PREFACE

The problem of material strength is one of the oldest and most important problem in technology. An important consideration for the choice of the load-bearing material and design in engineering applications is that the material should not fail while in service. In order to extract the benefit of the technological advances in material science, the engineering designer has to know when and why materials fail, under what kind of (extreme) stress conditions. This is true for their mechanical properties (fracture), electrical properties (dielectric and fuse breakdown) and other chemical properties (like corrosion or cavitation).

In this book, we try to present the basic concepts in order to understand the mechanical and electrical failures of solid materials containing inherent defects or disorders. Our emphasis has been on the question 'why', rather than on the question 'when', and we concentrate mainly on the statistical aspects of their failure strength distribution.

Failure phenomena are rarely intrinsic properties of the materials; rather they are caused by the various kinds of defects in the sample occurring at different length scales. Here, we consider mainly two extreme cases of disorder: weak disorder and strong disorder near the percolation threshold. Although the first case appears to be most useful one for engineering purposes, we think the second case may be more instructive in understanding the failure. It is precisely in this case of strong disorders, that considerable advances have been made recently by the physicists, who came with a large number of very successful applications of statistical physics models and theories like the percolation theory and the theories of critical phenomena. We concentrate mostly on these aspects of failure statistics in this book.

Because of our interest in basic principles, rather than in the applications, we present a large number of computer simulations and laboratory simulations or 'table top' experiments. The benefit of this approach is obvious from its success in the recent investigations and the literature. In an introductory chapter in the proceedings of a recent meeting on similar topics in Calcutta in 1993, while pledging for this kind of approach, Etienne Guyon observed and remarked that this particular approach ' ... is often not taken seriously by engineering communities. It is the use of "toy" and "table top" experiments connected with simple numerical simulations. An adequate project of this type implies that you give the most crucial or pertinent parameters and that you develop adequate as rigorous as possible analogies ...' (*Nonlinearity and breakdown in soft condensed matter*, Lecture Notes in Physics, volume 437, Springer-Verlag, 1994).

This book presents three particular cases of failure: Chapter 2 is on electrical failures like the fuse and dielectric breakdown problems and Chapter 3 is on mechanical fracture, both essentially in static models of solids containing random defects. We start with the electrical failures, because it helps to introduce several crucial concepts perhaps more easily. The last chapter is devoted to the recent model studies of dynamic failures like the earthquakes. If we insist more on the statics, rather than on the dynamics, that is merely because the dynamic problems, being more complex, have yielded less to solutions. We introduce in Chapter 1 the general concepts that we have employed in the subsequent chapters. We have made some attempts not to make the chapters totally interdependent, and, unlike the well-organised people, we did not try to avoid some repetitions when we thought some repetitions might help smooth reading of the book.

The study of failure or breakdown phenomena is a very active field of research today, and no doubt much of the material of this book has to be rewritten soon. Although we have made every effort to introduce all the important investigations and findings in this rapidly developing field, we apologise in advance if some important contributions are overlooked. We have tried to convey the spirit of the present-day research in this field, and enthuse the young researchers to tread new paths! If we have failed in our effort, then, not unlike the causes of material failures discussed here, only our defects and mistakes are to be blamed.

We are extremely grateful to our colleagues Muktish Acharyya, David Bergman, Debashish Chowdhury, Subhrangshu Sekhar Manna, Purusattam Ray, Pinhas Ron, Apurba Roy and Dietrich Stauffer, with whom we have studied and developed some of these models, experiments and theories. During the last ten/twelve years of the development of these studies, we have been benefited from various discussions with, and communications, comments and suggestions from Muktish Acharyya, Garani Ananthakrishna, Kamal Bardhan, David Bergman, Amit Dutta, Phil Duxbury, Paul Leath, Alex Hansen, Hans Herrmann, Subharngshu Sekhar Manna, Arkojyoti Misra, Purusattam Ray, Stephane Roux, Muhammad Sahimi, Asok Sen, Parongama Sen, Dietrich Stauffer, Robin Stinchcombe and Julia Yeomans. We are extremly thankful to Roger Elliott, for his encouragement and cooperation in connection with the publication of this book. Also, precisely in the above mentioned Calcutta meeting, we hatched the idea of a collaboration which took the form of writing this book, completed in another recent meeting this year again in Calcutta. We are grateful to Technion-Israel Institute of Technology (through Technion V. P. R. Fund), Haifa, and the Saha Institute of Nuclear Physics, Calcutta, for making this work and the visits possible. We are also grateful to the American Physical Society, Maryland, Elsevier Science, Amsterdam, Les Editions de Physique, Paris, and Springer-Verlag, Heidelberg, for their kind permissions to reproduce some of the figures (in part or full) from the journals and/or books published by them.

B. K. C.
L. G. B.

Calcutta and Haifa
December 1996

CONTENTS

1 Introduction   1
   1.1 Breakdown of disordered solids and dynamic frictional failures   1
   1.2 A brief introduction to some theoretical ideas and models   4
     1.2.1 A brief summary of percolation theory   5
     1.2.2 Stress concentration and statistics of extremes   20
     1.2.3 Self-organised criticality and sandpile models   27

2 Electrical breakdown in disordered solids   30
   2.1 Introduction   30
   2.2 The fuse problem   33
     2.2.1 Qualitative analysis   33
     2.2.2 Quantitative analysis: most probable failure current and distibution   36
     2.2.3 Numerical simulations and experimental results   45
     2.2.4 Other kinds of disorder: distribution of the failure threshold   48
     2.2.5 The shortest path and the electromigration fuse model   52
     2.2.6 Effects of temperature, AC fields and nonlinearity   56
   2.3 The dielectric breakdown problem   61
     2.3.1 Qualitative analysis   61
     2.3.2 Duality in two dimensions   61
     2.3.3 Lattice percolation   64
     2.3.4 Continuum percolation   67
     2.3.5 The shortest path   67
     2.3.6 Dielectric breakdown with tunnelling bonds   68
     2.3.7 Numerical simulations and experimental results   70
   2.4 Conclusions   78

3 Fracture strength of disordered solids — 80
- 3.1 Introduction — 80
- 3.2 Fracture strength of a perfect solid containing a single crack — 82
  - 3.2.1 Stress concentration — 84
  - 3.2.2 Griffith's energy balance concept — 86
  - 3.2.3 Experimental and computer simulational verifications of Griffith's law — 88
- 3.3 Fracture strength of brittle solids with small disorder and rough cracks — 91
  - 3.3.1 Griffith's law for fractal crack surfaces — 92
  - 3.3.2 Experimental observations — 94
- 3.4 Fracture strength of strongly disordered solids — 95
  - 3.4.1 Estimates for the fracture exponent and comparisons with experiment — 96
  - 3.4.2 Brittleness and plastic yield — 105
- 3.5 Fracture strength distribution — 106
  - 3.5.1 Extreme statistics and strength distribution — 106
  - 3.5.2 Comparisons with computer simulational and experimental results — 109
- 3.6 Fracture strength scaling in systems with random breaking thresholds — 113
  - 3.6.1 Models with random spring constants — 114
  - 3.6.2 Models with random breaking strengths — 114
- 3.7 Dynamics of fracture — 117
  - 3.7.1 Fracture propagation velocity — 117
  - 3.7.2 Large propagation velocity and morphology of fractured surfaces — 119
  - 3.7.3 Elastic precursor effects of complete fracture — 121
- 3.8 Dynamic annealed impurity and self-organised criticality in fracture — 126
- 3.9 Summary and conclusions — 127

4 Earthquakes in model systems — 128
- 4.1 Introduction — 128
- 4.2 Burridge-Knopoff stick-slip model of earthquakes — 130
  - 4.2.1 Laboratory simulation model — 130
  - 4.2.2 Computer simulation model — 133
  - 4.2.3 Numerical studies and results — 135
- 4.3 Self-organised criticality and cellular automata models of earthquakes — 140
- 4.4 Earthquake fault patterns and percolation model of earthquakes — 143

4.5	Precursors of self-organised criticality and earthquakes	145
	4.5.1 Pulse response of the sandpile model	146
	4.5.2 Response of the Burridge-Knopoff model to localised periodic pulses	148
4.6	Summary and conclusions	149

References   150

Index   159

1

INTRODUCTION

1.1 Breakdown properties of disordered solids and dynamic frictional failures

If one applies tensile stress on a solid, the solid elongates and gets strained. The stress (σ) - strain (ϵ) relation is linear for small stresses (Hooke's law) after which nonlinearity appears, in some cases. Finally at a critical stress σ_f, depending on the material, amount of disorder and the specimen size etc., the solid breaks into pieces; fracture occurs. In the case of brittle solids, the fracture occurs immediately after the Hookean linear region, and consequently the linear elastic theory can be applied to study the essentially nonlinear and irreversible static fracture properties of brittle solids (Lawn and Wilshaw 1975, Thomson 1986, Evans and Zok 1986).

Similarly, if one applies a (DC) voltage V across a conducting electrical circuit, a current I flows through it. For small voltages, the $I - V$ relation is linear (Ohm's law). Finally, at a critical current density I_f the circuit fuses as the current through some part of the circuit exceeds its threshold value. In dielectric materials, when the voltage gradient or the electric field exceeds its threshold value E_b, a similar (dielectric) breakdown occurs. These breakdowns occur, and can consequently be modelled, at classical or semiclassical levels. However, macroscopic breakdown can also occur due to the microscopic (quantum mechanical) tunnelling, as in Zener-type breakdown in semiconductors (Duke 1969).

With extreme perturbation, therefore, the mechanical or electrical properties of solids tend to get destabilised and failure or breakdown occurs. In fact, these instabilities in the solids often nucleate around disorder, which then plays a major role in the breakdown properties of the solids. The growth of these nucleating centres, in turn, depends on various statistical properties of the disorder, namely the scaling properties of percolating structures, its fractal dimensions, etc. These statistical properties of disorder induce some scaling behaviour for the breakdown of the disordered solids. Recently, there have been considerable developments in such studies; see for example, Englman and Jeager (1986), Chakrabarti (1988), Herrmann and Roux (1990), Charmet et al. (1990), Bergman and Stroud (1992), Yagil et al. (1993), Bardhan et al. (1994), and Sahimi (1997) for recent reviews.

Obviously with more and more random voids, the linear response of e.g. the modulus of elasticity Y (say, the Young's modulus) of the solid decreases, or with more and more nonconducting elements in a conducting network, the conductivity Σ of the network decreases and with more and more conducting elements in a dielectric, its inverse dielectric constant decreases. So also does the breaking strength of the material: the fracture strength σ_f of the specimen, or the fuse current I_f of the network, or the dielectric breakdown voltage V_b of the dielectric (or the Zener breakdown voltage of impure insulators or semiconductors), all decrease on the average with the increased concentration of random impurities (random voids, insulators or conductors respectively). For studying most of these mechanical (elastic) or electrical breakdown problems of randomly disordered solids, we take the lattice model of elastic or conducting disordered solids. In these lattice models, a fraction p of the bonds (or sites) are intact, with the rest $(1-p)$ being randomly broken or cut (for random mechanical networks), or a random fraction p of the bonds (or sites) are conductors, with the rest being insulators (for random electrical networks). Fluctuations in the random distribution give rise to random clusters of impurities inside the bulk, for which the (percolation cluster) statistics is well developed (Stauffer and Aharony 1992), and we investigate here the effect of these voids or impurity clusters on the ultimate strength of the bulk solid. We also consider and compare the results for failure strength of the solids with random bond strength distribution, where the solids do not have any particular void cluster as the local strength varies continuously and the percolation cluster statistics can not be applied directly in such cases.

As is well known, the initial variations (decreases) of the linear responses like the elastic constant Y and conductivity Σ and also the breakdown strengths σ_f, I_f or V_b are analytic with the impurity (dilution) concentration. Near the percolation threshold (see next section) p_c, up to (and at) which the solid network is marginally connected through the nearest neighbour occupied bonds or sites and below which the macroscopic connection ceases, the variations in these quantities with p are expected to become singular; the leading singularities being expressed by the respective critical exponents. The exponents (T_e and t_c respectively) for the modulus of elasticity $Y \sim (\Delta p)^{T_e}$, with $\Delta p = (p - p_c)/p_c$, and conductivity $\Sigma \sim (\Delta p)^{t_c}$ for $p > p_c$, are well-known and depend essentially on the dimension d of the system (see next section). Our interest here is to find the corresponding singularities for the essentially nonlinear and irreversible properties of the mechanical and electrical breakdown strengths for p near the percolation threshold p_c: to find the exponents (T_f, t_f and t_b respectively) for the average fracture stress $\sigma_f \sim (p - p_c)^{T_f}$, fuse current $I_f \sim (p - p_c)^{t_f}$ and dielectric breakdown voltage or field V_b or $E_b \sim (p_c - p)^{t_b}$ of randomly diluted networks near p_c. Very often, one maps the problem of breakdown to the corresponding linear problem (assuming brittleness or ohmic behaviour up to the breaking point) and then derives the scaling relations giving the breakdown exponents (T_f, t_f and t_b) in terms of the linear response exponents (T_e or t_c) and other lattice statistical exponents.

Unlike that for the 'classical' linear responses of such solids, the extreme nature of the breakdown statistics, nucleating from the weakest point of the sample, gives rise to a non-self-averaging property. We will discuss these distribution functions $F(\sigma)$, or $F(I)$, or $F(E)$, giving the cumulative probability of failure of a disordered sample of linear size L. We show that the generic form of the function $F(\sigma)$ can be either the Weibull (1951) form

$$F(\sigma) \sim 1 - \exp\left[-L^d \left(\frac{\sigma}{\Lambda(p)}\right)^{1/\psi}\right],$$

or the Gumbel (1958) form

$$F(\sigma) \sim 1 - \exp\left[-L^d \exp\left(-\frac{\Lambda(p)}{\sigma}\right)^{1/\psi}\right],$$

where $\Lambda(p)$ is determined by the linear responses (like elasticity or conductivity) of the disordered solid and by some other lattice statistical quantities etc., and ψ is an exponent discussed later. A finite value of $F(\sigma)$ at the most probable strength σ_f then induces a nontrivial system size dependence for strengths σ_f, I_f or V_b, apart from the singular variations with disorder concentrations (p near p_c) through $\Lambda(p)$, determined by the exponents like T_f, t_f or t_b. We emphasise in this review all such scaling relations, whenever possible. The available experimental verifications and comparisons of such scaling relations, in computer simulations and in various laboratory simulated random networks, are discussed.

Also, many solid structures, like portions of the earth's crust, get dynamically destabilised due to the slow movements of the tectonic plates on which the crust-portion rests, and the elastic forces coming from the neighbouring portions of the crust. Here, the elastic energy grows due to the strain or relative displacement of the moving portion of the earth's crust, until it reaches a 'self-organised' level when the (dynamic frictional) instabilities are settled through a series of energy releases (E_r) or 'earthquakes'. Such dynamic failure models have attracted recently a lot of interest, and considerable advances have been made in such studies; see for example, Carlson et al. (1994) and Perez et al. (1996) for recent reviews. These studies mainly attempt to capture the Guttenberg-Richter type power law (Guttenberg and Richter 1954) for the cumulative frequency $n(E_r)$ of the energy releases: $n(E_r) \sim E_r^{-c}$, where $c \simeq 0.8$ to 1.1 from the recorded data of the earthquake events.

In the rest of this chapter, we will discuss briefly the theoretical ideas and the models employed for the study of failure of disordered solids, and other dynamical systems. In particular, we give a very brief summary of the percolation theory and the models (both lattice and continuum). The various lattice statistical exponents and the (fractal) dimensions are introduced here. We then give brief introduction to the concept of stress concentration around a sharp edge of a void or impurity cluster in a stressed solid. The concept is then extended to derive the extreme statistics of failure of randomly disordered solids. Here, we also discuss the competition between the percolation and the extreme statistics in determining the breakdown statistics of disordered solids. Finally, we discuss the self-organised criticality and some models showing such critical behaviour.

In the next chapter (Chapter 2), we estimate the fuse current of a conducting random network or the breakdown field of a randomly metal-loaded dielectric, using the percolation cluster models and their statistics. We also discuss here the breakdown probability distributions of such networks. All these theoretical estimates are compared with the extensive experimental and computer simulation results.

We shall discuss in Chapter 3 the scaling properties of fracture of brittle percolating solids. For a brittle solid with a single linear crack inside, Griffith (1920) obtained the breakdown strength or the fracture stress by equating the elastic energy, released due to extension of the micro crack, with the surface energy of the newly grown surface of the crack. This procedure, therefore, maps the nonlinear (and irreversible) breakdown problem to the linear (and reversible) elastic problem of brittle systems. This idea has been extended to percolating solids, where the pre-existing cracks due to random vacancies or voids are of different sizes and shapes governed by the percolation statistics. The resulting scaling properties and the statistics of fracture strength are now quite well studied and established. We give here a brief account of these developments, and compare the theoretical result with the experimental, laboratory or computer simulation results.

We discuss the various dynamical models of earthquake-like failures in Chapter 4. Specifically, the properties of the Burridge-Knopoff stick-slip model (Burridge and Knopoff 1967) and of the self-organised criticality models, the Guttenberg-Richter type power laws, for the frequency distribution of 'earthquakes' in these models are discussed here.

1.2 A brief introduction to some theoretical ideas and models

We give here a brief discussion of the various theoretical models and concepts or ideas, employed later for the understanding of the failure properties of disordered solids. We consider here mainly the quenched disordered solids, where the defects do not change their locations with the growth or propagation of the failure or the breakdown. In particular, we discuss here the percolation models and the theory, the concepts of stress concentration and of extreme statistics and of the self-organised criticality. As mentioned above, the theoretical ideas and models discussed here form the basis for modelling and analysing the investigations and the results of various kinds of failures discussed in the following chapters.

In the first reading of this book, a detailed study of this section is not essential, although a general familiarity with the concepts and the notations is necessary.

1.2.1 A brief summary of percolation theory

As mentioned before, the disordered solids will be mostly modelled in this book using randomly diluted site or bond lattice models. A knowledge of percolation cluster statistics will therefore be necessary and widely employed. Although this lattice percolation kind of disorder will not be the only kind of disorder used to model such solids, as can be seen later in this book, the widely established results for percolation statistics have been employed successsfully to understand and formulate analytically various breakdown properties of disordered solids. We therefore give here a very brief introduction to the percolation theory. For details, see the book by Stauffer and Aharony (1992).

Consider a nonconducting plate, on which one may spray uniformly some conducting dye. If one applies a small potential difference across any two opposite ends of the plate, with an ammeter in series, one obviously does not get any current initially when no conducting dye is sprayed on the plate. Eventually when the entire plate is painted with the dye, current would flow through the surface network of conducting dye. However, this transition from vanishing current to a nonvanishing current through the surface dye network does not occur at 100 per cent surface coverage by the paint. Rather, it occurs at the percolation transition point where the overlapping clusters of dye grains (assumed to be sufficiently large, almost macroscopic, in size and, say, disk-shaped) just percolate, or form a marginally connected path due to fluctuations in the random distribution, across the plate. Until the percolation point, the dye-clusters (formed out of overlapping dye disks) are finite and do not contribute to any (classical) macroscopic property or transport-type responses, which start to manifest above the percolation point. Once this marginally connected path or the macroscopic network is formed, spraying more paint would contribute, by connecting more of the finite clusters to the infinite or percolating one, to the net conductivity of the percolating network. The conductivity of the

6   Introduction

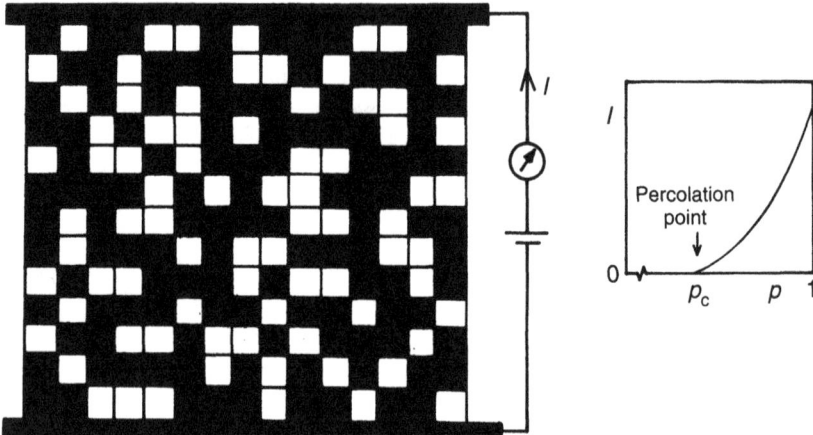

FIG. 1.1. A random conducting network with the conducting blocks (denoted by black squares) with concentration above the percolation threshold. If one assumes the conducting clusters to be formed when the blocks are connected by the nearest-neighbour sites (not by the marginally touching corners), the percolation problem is a random site problem. The current I through the network decreases to zero if the conducting block concentration p falls below the percolation threshold p_c, as shown in the figure on the right side.

network is thus zero until a percolation threshold coverage of the surface by the conducting dye, and it starts growing from this vanishing value as the average conductor concentration grows beyond the percolation threshold.

Similarly, one can study the growth of the elastic constants (say the rigidity modulus) of a randomly formed elastic network, near the percolation point. The central force elastic problem (for networks formed out of linear springs only) belongs however to a different class of percolation problem, known as elastic percolation or central force percolation, and is discussed separately later (see Section 1.2.1(f)).

The growth of the conductivity or elasticity of such networks near their respective percolation threshold points can be expressed as powers (known as exponents) of the interval (of random concentration of the conducting or elastic material) from the percolation threshold. These powers or the exponents are observed to be universal, in the sense that they do not depend on many details of the problem or of the lattice, but depend on only some subtle geometric features of the problem; e.g., the exponents often depend only on the lattice dimensionality.

In order to make the discussion more quantitative and precise, let us

A brief introduction to some theoretical ideas and models 7

Table 1.1 *The site and bond percolation thresholds for different lattice types*

Lattice	Site	Bond
Square	0.59275	1/2
Triangular	1/2	$2\sin(\pi/18) \simeq 0.34729$
Honeycomb	0.6962	$1 - 2\sin(\pi/18) \simeq 0.65271$
Diamond	0.428	0.388
Simple cubic	0.3116	0.2488
BCC	0.246	0.1803
FCC	0.198	0.119

consider now the lattice percolation model. Here, it can have two versions: site percolation and bond percolation. In the site percolation problem, each site of a large (infinite) lattice is occupied randomly with a probability p, independent of its neighbours. A random site lattice version of the conducting random electrical network, discussed earlier, is shown in Fig. 1.1. Clusters are defined as the graph of neighbouring occupied sites. The term 'neighbouring' is defined appropriately for the problem. Most often, as in Fig. 1.1, the percolation connectivities occur through nearest neighbours and hence the 'neighbours' in the cluster definition mean nearest neighbours. In bond percolation problems, each bond of the lattice is occupied with the random probability p, and the clusters are defined as the graphs of overlapping bonds, sharing common sites. As discussed earlier, most of the physical properties of such random systems depend in an essential way on the geometric properties of these random clusters, and in particular, on the existence of an infinite connected cluster which spans the system. Percolation theory deals with the statistics of the clusters formed.

In random percolation problems, there exists a unique threshold concentration p_c, at which a lattice spanning cluster, connecting opposite ends of the (very) large lattice, is formed. This threshold concentration p_c, also called the percolation probability, depends on the lattice and the type of percolation problem. The percolation thresholds for some simple lattices are given in Table 1.1. These percolation thresholds p_c for site and bond problems are topologically related and it can be shown rigorously that p_c for site percolation is higher or equal to the p_c for bond percolation. At least one percolating or lattice spanning cluster is statistically formed at and above the percolation threshold. More than one lattice spanning cluster spanning in all the lattice directions is not permitted by the topology in two dimensions; however, more than one such lattice spanning clusters are geometrically possible in higher dimensions, and are now known to exist. When one considers spanning in one direction only, one can have more than one spanning cluster even in two dimensions (see e.g. Sen 1996 and references therein).

(a) Cluster statistics

We now define some statistical quantities of interest in percolation theory. Let $n_s(p)$ denote the number of clusters (per lattice site) of size s. In fact, a detailed knowledge of $n_s(p)$ would give us a lot of information on the percolation statistics, as most of the quantities of interest can be extracted from various moments of the cluster size distribution n_s. Although, in general, we do not have any analytic knowledge about this distribution function $n_s(p)$ near p_c, we can utilise the powerful observation of scaling behaviour of $n_s(p)$ near p_c (see the next section).

The probability that a given site (bond) is occupied and is a part of an s-size cluster is $sn_s(p)$. Let us denote by $P(p)$, the probability that any occupied site (bond) belongs to the infinite (lattice spanning) cluster. Then we have the obvious relation,

$$\sum_s sn_s + P = 1, \tag{1.1}$$

where the summation extends over all finite clusters. Clearly, at $p = 1$, $P(p) = 1$, and $P(p) = 0$ for $p < p_c$ as the infinite cluster does not exist for $p < p_c$. $P(p)$ can therefore be taken as the order parameter for this geometric (percolation) phase transition. Another quantity of interest is the average mean size of the finite clusters, denoted by $S(p)$, which is related to $n_s(p)$ through the relation

$$S(p) = \frac{\sum_s s^2 n_s(p)}{\sum_s sn_s(p)}, \tag{1.2}$$

since the denominator remains finite at p_c, and the summation is again over all the finite clusters.

One also defines a pair connectedness function $C(p, r)$ as the probability that two occupied sites (bonds) at a distance r are members of the same cluster. In fact, integration over the pair connectedness over all distances gives the mean cluster size: $S(p) = \sum_r C(p, r)$.

(b) Critical exponents

As mentioned earlier, near p_c, most of the quantities defined above have power law variations. For example, the variation of the total number of finite clusters per site $G(p) = \sum_s n_s(p)$ (with the summation over all finite clusters), the decay of the order parameter $P(p)$ and the divergence of the mean cluster size $S(p)$, as $p \to p_c$, can be expressed by power law variations of these quantities with the concentration interval $|p - p_c|$ as

Table 1.2 *Values of the percolation critical exponents in various dimensions*

Exponent	$d = 2$	$d = 3$	$d \geq 6$ (Bethe lattice)
α	$-2/3$	-0.62	-1
β	$5/36$	0.41	1
γ	$43/18$	1.80	1
η	$5/24$	$\simeq 0$	0
ν	$4/3$	0.88	$1/2$

$$G(p) \equiv \sum_s n_s(p) \sim |p - p_c|^{2-\alpha}, \tag{1.3a}$$

$$P(p) \sim (p - p_c)^\beta, \tag{1.3b}$$

$$S(p) \sim |p - p_c|^{-\gamma}, \tag{1.3c}$$

and

$$C(p, r) \sim \frac{\exp(-r/\xi(p))}{r^{d-2+\eta}}, \tag{1.3d}$$

where the correlation length

$$\xi(p) \sim |p - p_c|^{-\nu} \tag{1.3e}$$

diverges at $p = p_c$.

These powers α, β, γ, η and ν are called the critical exponents. These exponents are observed to be universal in the sense that although p_c depends on the details of the models or lattice considered, these exponents depend the only on the lattice dimensionality (see Table 1.2). It is also observed that these exponent values converge to the mean field values (obtained for the loopless Bethe lattice) for lattice dimensions at and above six. This suggests the upper critical dimension for percolation to be six.

(c) Scaling theory

Scaling theory assumes (Stauffer 1979, Stauffer and Aharony 1992) that the cluster distribution function $n_s(p)$ is a homogeneous one near $p = p_c$. Thus $n_s(p)$ is basically a function of the single scaled variable $s/S_\xi(p)$, where the S_ξ denote the typical cluster size:

$$n_s(p) \sim s^{-\tau} f\left(\frac{s}{S_\xi(p)}\right), \tag{1.4}$$

with $S_\xi(p) \sim |p - p_c|^{-1/\sigma}$. Here, τ and σ are two independent exponents and the scaling theory intends to relate all the above exponents (α, β,

γ, η and ν) to these exponents through the scaling relations. The scaling function $f(x)$ is also assumed to be asymptotically defined: $f(x) \to 1$ and $f(x) \to 0$ as $x \to 0$ and $x \to \infty$ respectively. The rest of the details remain unspecified in this theory. It may be noted that the above scaling form (1.4) gives $n_s(p_c) \sim s^{-\tau}$. This has been checked using Monte Carlo simulations very accurately (Stauffer and Aharony 1992).

Scaling Relations Assuming the above scaling form for $n_s(p)$, the m-th moment of $n_s(p)$ can be expressed as

$$\sum_s s^m n_s \sim \sum_s s^{m-\tau} f\left(\frac{s}{|p-p_c|^{1/\sigma}}\right)$$

$$\sim |p-p_c|^{\frac{\tau-m-1}{\sigma}} \int x^{m-\tau} f(x) \mathrm{d}x$$

$$\sim |p-p_c|^{\frac{\tau-m-1}{\sigma}},$$

assuming the integral over x ($= s/|p-p_c|^{\frac{1}{\sigma}}$) to converge because of the asymptotic behaviour of $f(x)$. Noting that $G(p), P(p)$ and $S(p)$ correspond to the zeroth, first and second moments of n_s, we get $\alpha = 2 - \frac{\tau-1}{\sigma}, \beta = \frac{\tau-2}{\sigma}$ and $\gamma = -\frac{\tau-3}{\sigma}$. This gives the scaling relation

$$\alpha + 2\beta + \gamma = 2, \tag{1.5}$$

satisfied by the observed values of the critical exponents (see Table 1.2). Also, since $S(p) = \sum_r C(r,p) = \int r^{d-1} C(r,p) \mathrm{d}r$, one immediately gets

$$S(p) \sim |p-p_c|^{-\gamma} = \int r^{d-1} \frac{e^{-r/\xi}}{r^{d-2+\eta}} \mathrm{d}r \sim \xi^{2-\eta} \int y^{1-\eta} e^{-y} \mathrm{d}y \sim |p-p_c|^{-\nu(2-\eta)},$$

giving the scaling relation

$$\gamma = \nu(2-\eta). \tag{1.6}$$

Assuming additionally that near p_c, the typical density $sn_s \sim S_\xi^{1-\tau}$ where $S_\xi \sim |p-p_c|^{-1/\sigma}$, one gets $sn_s \sim |p-p_c|^{(\tau-1)/\sigma}$ for the density. Asuming further that this density scales with the inverse of the typical volume element $\xi^d \sim |p-p_c|^{-d\nu}$ we get the hyper-scaling relation

$$d\nu = (\tau-1)/\sigma = 2 - \alpha. \tag{1.7}$$

These scaling relations are satisfied by the numerically estimated values, and in special cases the exact values, of the exponents given in Table 1.2. For the general theoretical estimates of the values of these exponents, for

A brief introduction to some theoretical ideas and models 11

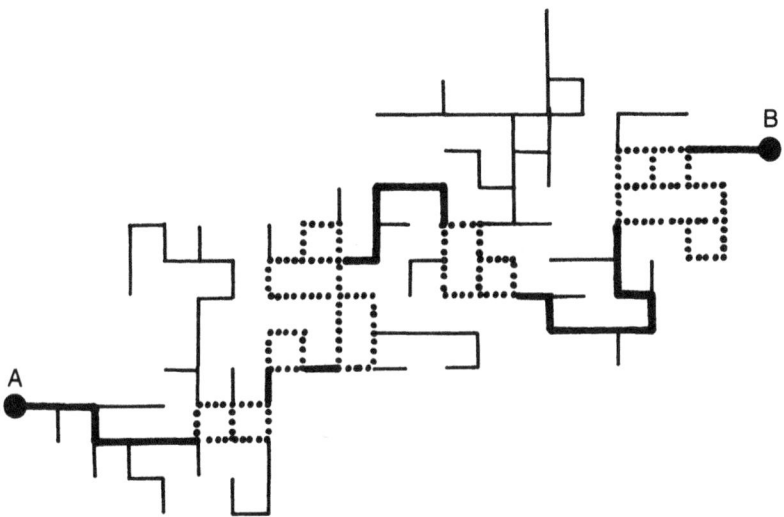

FIG. 1.2. Portion of a random bond percolating cluster backbone, connecting the points A and B. Here, the thick black lines represent the singly connected bonds or 'red' bonds which, if cut, will disconnect the connection between A and B. The bonds in the blob portions are indicated by dotted lines. The dangling bonds are indicated by thin black lines (cf. Stauffer and Aharony 1992).

example from the renormalisation group theory, see Stauffer and Aharony (1992) and the references therein.

(d) Structure of the spanning (or infinite) cluster and self-similarity

Since the infinite or the lattice spanning cluster allows for all the macroscopic transport (like conductivity) and other elastic, etc. properties of the network, the structure of this cluster at or very near the percolation threshold is of particular importance. Typically, such cluster consists of a backbone, which essentially is responsible for the above transport etc. properties of the networks. Additionally, there exist some dangling ends attached to this backbone. For classical transport, these dangling ends of of the percolating cluster are unimportant, and the backbone of the cluster is all-important. For example, Fig. 1.2 shows the picture of a portion of a current-carrying backbone of a random bond conductor-insulator network in two dimensions. Following Stanley (1977), the bonds of the backbone can be divided into two groups. In one group, the bonds are in the 'blobs', which are multiply connected, and in the other group one has the 'red bonds', which are such that, unlike those in blobs, if any is cut, the back-

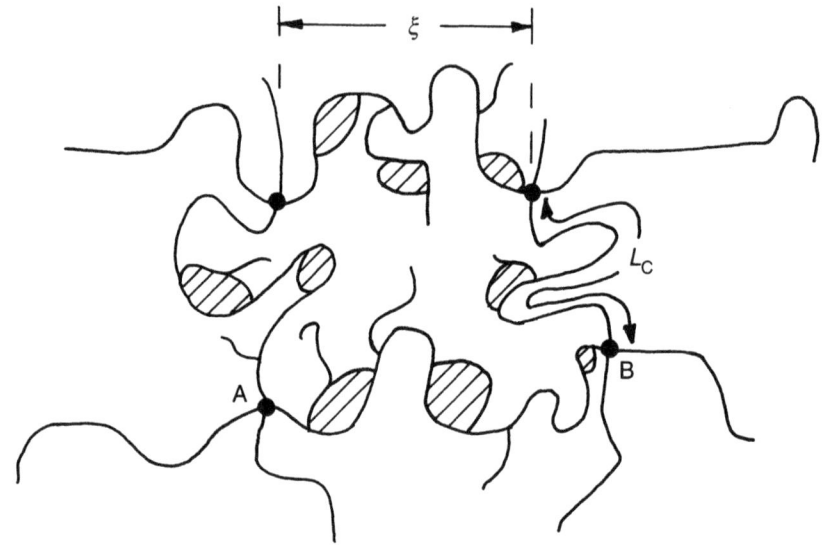

FIG. 1.3. A portion of the node-link-blob superlattice model for $p > p_c$. The distance between two nodes of the lattice (say A and B) is ξ, while the 'chemical' length of the 'tortuous' bond of the superlattice is L_c.

bone is split into two parts. The red bonds are also called the 'singly connected bonds' of the infinite cluster. In a current-carrying backbone, such red bonds carry the entire current through the network, while the current gets split among the multiple connections within the blobs.

In fact, if one measures the total number of bonds (sites) on the infinite cluster at the percolation threshold (p_c) in a (large) box of linear size L, then this number or the mass of the infinite cluster will be seen to scale with L as $L^{d_{IC}}$, where d_{IC} ($\leq d$) is called the fractal dimension of the infinite cluster at the percolation threshold. Similar measurements for the backbone (excluding the dangling ends of the infinite cluster) give the backbone mass scaling as L^{d_B}, $d_B \leq d_{IC}$, where d_B is called the backbone (fractal) dimension. In fact, d_{IC} can be very easily related to the embedding Euclidean dimension d of the cluster by

$$d_{IC} = d - \frac{\beta}{\nu}, \qquad (1.8)$$

because the mass of the infinite cluster is given by that occurring with probability P within a volume ξ^d, giving the infinite cluster mass $\sim \xi^{d_{IC}} \sim \xi^d P \sim \xi^d |p - p_c|^\beta \sim \xi^{d-\beta/\nu}$.

A brief introduction to some theoretical ideas and models

Table 1.3 *Fractal dimensions of the percolation clusters*

Fractal dimension	$d = 2$	$d = 3$	$d = 6$ (Bethe lattice)
d_{IC}	91/48	2.53	4
d_B	1.6	1.7	2
d_{min}	1.13	1.34	2

The node-link-blob model Very near the percolation threshold p_c, one can use a very elegant picture of the backbone cluster, suggested by Skal and Shklovskii (1974), de Gennes (1976) and Stanley (1977), known as the nodes-links-blobs picture. It is based on the idea that at length scales larger than the percolation correlation length ξ, the backbone is homogeneous and appears as a 'super-lattice', while for length scales less than ξ, the backbone is a fractal (see Fig. 1.3). The super-lattice is assumed to be formed of tortuous 'bonds' having the end-to-end length $\xi \sim |p - p_c|^{-\nu}$ and having the tortuous chain length or chemical length $L_c \sim |p - p_c|^{-\zeta}$. It may be noted that this chemical length L_c passes through the blobs and therefore its length depends on the property measured. For example, for electrical conductivity, the effective value of L_c would be different from that corresponding to the minimum or shortest path. The values of the fractal dimensions d_{IC}, d_B and $d_{min} = \zeta_{min}/\nu$ for the minimum chemical path are given are Table 1.3. It may be mentioned here that if L_c refers to the number of singly connected bonds (or sites) of the percolating backbone, then $\zeta = 1$ in all dimensions greater than unity (Stauffer and Aharony 1992).

In the above node-link-blob picture, the percolation cluster is self-similar up to a length scale ξ in the sense that starting from the length scale ξ, the links contain blobs (and the dangling ends) which, in turn, are composed of links and blobs (and the dangling ends) up to the lowest scale (of the lattice). This self-similarity extends up to infinite scale at the percolation threshold (where ξ becomes infinitely large).

In another picture of the infinite cluster at the percolation threshold, this self-similarity property gets built-in, using a carpet-like fractal. For example, at the zeroth level, one takes a square of linear size 1, and then at the next level constructs a bigger square of linear size 3×3 with the central square of size 1×1 removed. In the next stage, when the linear size becomes 9×9, the central square of size 3×3 is again completely removed, as shown in Fig. 1.4. Here these three stages of constructing the fractal are shown. This process can be continued further and one gets the Sierpinski carpet in the infinite limit of the process. One can easily check that the mass M of such a fractal scales with its linear size L as $M \sim L^{d_f}$, with $d_f = \ln 8 / \ln 3$. This is because if $L = 3^n$, then $M = 8^n$ here, because of the construction. It may be noted that instead of the above deterministic

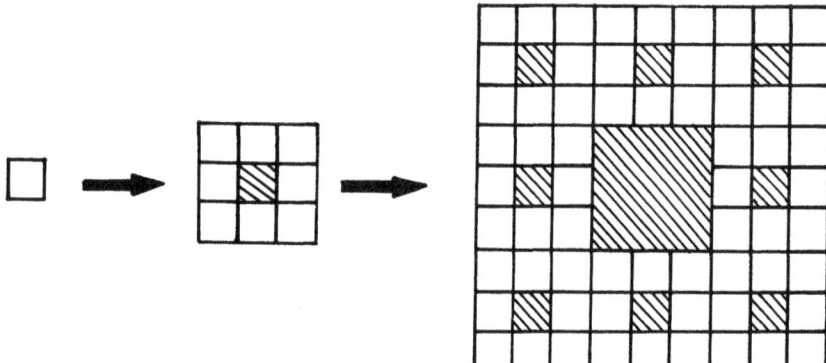

FIG. 1.4. The successive initial stages of constructing a Sierpinsky carpet. The fractal is obtained at the infinite limit of this construction process. The empty squares are shaded. At each step of iteration, the linear dimension of the carpet is increased by a factor 3, while its mass increases by a factor 8.

way of constructing the fractal, where the central block is always removed, one could choose a random process of removing the same size of block, but from arbitrary regions. This would result in a random fractal with the same fractal dimension. It may also be noted that this fractal dimension $d_f = \ln 8/\ln 3 \simeq 1.893$ is indeed close to the $d_{IC} = 91/48 \simeq 1.896$ of the infinite cluster at p_c in $d = 2$. A straightforward generalisation in three dimensions, where 7 cubes are empty for each 27 cubes in scale $L = 3$, gives $d_f = \ln 20/\ln 3 \sim 2.73$, which is again close to $d_{IC} \simeq 2.53$.

(e) Conductivity of percolating networks

If we consider a random bond network, where the bonds are conductors with concentration p and insulators with concentration $1 - p$, then such a network has a macroscopic conductivity $\Sigma(p)$ (measured by the ratio of the net current across the two ends of the network to the voltage across it) for $p \geq p_c$, the percolation threshold for the lattice. Obviously $\Sigma(p) = 0$ for $p < p_c$, and one observes the conductivity $\Sigma(p)$ to grow with p above p_c following a power law

$$\Sigma(p) \sim (p - p_c)^{t_c}, \tag{1.9}$$

where t_c is called the conductivity exponent. Extensive numerical studies give (see Stauffer and Aharony 1992) $t_c \simeq 1.3, 2.0$ and 3 in $d = 2, 3$ and 6 (Bethe lattice) respectively.

One can similarly consider a random network of superconducting bonds with infinite conductance, and conducting bonds with finite (say, unit) conductance. If the superconductor concentration p is above p_c, then the network conductivity $\Sigma(p)$ is infinity, while it becomes finite for p below p_c. For p near p_c, $\Sigma(p)$ diverges following a power law

$$\Sigma(p) \sim (p_c - p)^{-s_c}, \qquad (1.10)$$

where s_c is the superconductivity exponent of the network. Computer simulation studies gave (see Stauffer and Aharony 1992) $s_c \simeq 1.3$, 0.7 and 0 in $d = 2$, 3 and 6 (Bethe lattice) respectively (Bunde et al 1985). Several experimental realisations of such random conductor-insulator mixtures and superconductor-conductor mixtures or networks exist, and the experimental results for t_c and s_c compare well with the above numerical estimates.

Several attempts have been made to relate these conductivity exponents t_c or s_c with the percolation cluster statistical exponents discussed earlier. For example, in the node-link-blob model, the conductivity Σ of the network is proportional to the number $\sim \xi^{-(d-1)}$ of parallel links, while the conductivity of each link decreases with the number $\sim \xi^{-1}$ of the link-elements (between two successive nodes) and their effective resistance proportional to the chemical length $L_c \sim \xi^{\zeta/\nu}$. This gives $\Sigma \sim \xi^{-(d-2)-\zeta/\nu} \sim (p - p_c)^{t_c}$; $t_c = (d-2)\nu + \zeta$. Here, however, the measure of the chemical length depends on the effective resistance of the multiply connected blobs, which exists for $d < 6$. This makes the estimate of the exponent ζ ($= 1$ for $d \geq 6$) uncertain for $d < 6$, and does not allow one to estimate the above exponent t_c from the above scaling relation (except at $d = 6$, where ζ is known exactly). Several other attempts to relate the conductivity exponent with the percolation cluster statistical exponents have been made, and it appears (see Stauffer and Aharony 1992) that the conjecture $t_c = [(3d-4)\nu - \beta]/2$ fits quite well with the numerical results. Also, through duality transformation, the resistance of the random conductor-insulator network is equal to the conductivity of the random superconductor-conductor network on the dual lattice. This suggests $t_c = s_c$ in two dimensions. Generally, it seems (Stauffer and Aharony 1992) that the scaling conjecture $s_c = 2 - \alpha - t_c$ fits well with the numerical estimates of s_c and t_c.

(f) Elasticity of percolating networks and elastic percolation

Let us consider a random bond network with both central and bond-bending force constants. The elastic energy of the network is then contributed by both the central force part, because of the stretching of the bond length, and by the bond-bending part, because of the changes in the bond angles

$$H = \frac{\kappa_c}{2}\Sigma_{<i,j>}[(\mathbf{u}_i - \mathbf{u}_j) \cdot \mathbf{R}_{ij}]^2 p_{ij} + \frac{\kappa_b}{2}\Sigma_{<ijk>}(\delta\theta_{ijk})^2 p_{ij}p_{jk}, \qquad (1.11)$$

where $\mathbf{u}_i$ denotes the off-lattice displacement of the ith lattice site, $\mathbf{R}_{ij}$ the lattice separation vector between sites i and j, θ_{ijk} the angle at site j

between the bonds (ij) and (jk). Here κ_c and κ_b denote the central and bond-bending force constants respectively, and $p_{ij} = 1$ if the (ij) bond is present and $p_{ij} = 0$ otherwise. It is clear that for nonvanishing κ_b, the elastic restoring force or the elastic modulus (denoted generally by Y) of the network, both for longitudinal and shear strains, is zero for the random bond occupation concentration p below p_c and Y grows to nonvanishing values for $p > p_c$, because of the existence of at least one connected path. It is observed that Y grows following a power law

$$Y \sim (p - p_c)^{T_e} \tag{1.12}$$

for $p > p_c$, where T_e is called the elastic exponent. Extensive numerical studies give $T_e \simeq 3.96, 3.75$ and 4 in $d = 2, 3$ and 6 respectively (Stauffer and Aharony 1992, Sahimi 1995). Experimental results also compare well with the above-mentioned numerical estimates.

This elasticity of the random rigid bonds (with finite rigidity) and non-rigid bonds (considered as absent bonds) is analogous to the problem of conductivity of a random conductor bond (with finite conductivity) and insulator bond mixture, considered in the previous section. One can now also consider the elasticity of super-elastic networks (see Benguigui and Ron 1993) where, like a superconductor-normal conductor mixture (discussed in the previous section), the random bonds may be infinitely rigid occurring with probability p and finitely rigid (unit rigidity) with probability $1 - p$. For $p > p_c$ then the network elastic modulus Y is infinity, while it increases following the power law

$$Y \sim (p_c - p)^{-s_e}, \tag{1.13}$$

for $p \leq p_c$, where s_e is the superelastic exponent. Best numerical estimates for the exponents for the bond bending model are: $s_e \simeq 1.24, 0.56$ and 0 in $d = 2, 3$ and 6 respectively. These values also compare well with the experimental results (Benguigui and Ron 1993). It may be noted that, unlike the cases of conductor-insulator and superconductor-conductor mixtures, here the elasticity exponents T_e and s_e of random elastic networks and superelastic networks are not identical in $d = 2$.

Again several attempts have been made to relate the elastic exponents T_e and s_e with the conductivity exponents t_c and s_c and other percolation exponents. Identifying the elastic energy $E \sim F^2/Y \sim \Gamma^2/\xi^2 Y$ of the elastic network under stress F or torque $\Gamma \sim F\xi$ with the resistive loss $E \sim I^2/\Sigma$ of a random conductor-insulator network, and also identifying the torque Γ of the elastic network with the current I of the random resistor network, one gets (see e.g. Sahimi 1995) $\xi^2 Y \sim \Sigma$ or $T_e = t_c + 2\nu$. The numerical values of T_e and t_c, as given before, compare well with this scaling

A brief introduction to some theoretical ideas and models

Table 1.4 *Percolation thresholds and exponents for central force elastic percolation*

Dimension	Lattice	p_{ce}		Exponents		
		Site	Bond	ν^{ce}	β^{ce}	d_B^{ce}
2	square	1	1	1.16	0.19	1.78
	triangular	0.6975 ± 0.0003	0.6603 ± 0.0002	± 0.02	± 0.02	± 0.02
3	simple cubic	1	1	–	–	$\simeq 2.5$
	BCC	0.737 ± 0.002	–			

relation. It has also been conjectured that $s_e = \nu - \beta/2$ (Sahimi 1995), which again compares well with the above numerical estimates for s_e.

Central force elastic percolation If one considers an elastic network of springs, which can provide only the central force, and not the bond-bending force considered earlier, then the elastic energy of such a random bond network can be expressed by the same energy function H in (1.11) with $\kappa_b = 0$. It is clear that, both for site and bond percolation, the threshold concentration p_{ce} of the bonds (springs) beyond which the network can not sustain any longitudinal or shear stress will in general be greater than the (scalar) percolation threshold p_c discussed earlier. This is because the marginally connected cluster formed at p_c contains a finite fraction of linearly or singly connected bonds or springs joining at different angles (the fraction of red bonds in the node-link-blob model discussed earlier). For a central force network, such joints fail to provide any resistance, as the angle between the neighbouring bonds changes without stretching the bonds (springs). Higher occupation concentrations are required for development of the restoring force. For higher values of p, the higher connectivity of the percolating backbone can ensure that none of the bond angles is between two freely joined springs and can be changed freely. In particular, for hypercubic lattices in any dimension, like a square ($d = 2$) or simple cubic lattice (in $d = 3$), $p_{ce} = 1$ for both site and bond dilution. Extensive computer simulations give the central force percolation thresholds p_{ce} (see Moukarzel and Duxbury 1995, Jacobs and Thorpe 1995, Sahimi 1995) as given in Table 1.4. It may be noted that both the values of the thresholds p_{ce} as well as the exponents given in Table 1.4 are tentative (particularly in $d = 3$), and also there are controversial indications suggesting different universality classes (exponents) for bond and site central percolations (see e.g. Sahimi 1995).

The growth of the elastic modulus Y of such a random central force network with $p > p_{ce}$ again follows a power law $Y \sim (p - p_{ce})^{T_{ce}}$. The available estimates for $T_{ce}/\nu_{ce} \simeq 1.12 \pm 0.05$ in $d = 2$. Also, for superelastic percolation with central force, where a fraction p of the bonds (springs) are infinitely rigid and the rest are central force springs (with finite spring constant), the elastic modulus $Y \to \infty$ for $p > p_{ce}$, while $Y \sim (p_{ce} - p)^{-s_{ce}}$ for $p \leq p_{ce}$. The present estimate gives $s_{ce}/\nu_{ce} \simeq 0.92 \pm 0.02$ (see Sahimi 1995).

(g) Continuum percolation

So far, we have discussed discrete lattice percolation where, if a bond (or site) is occupied, it has a fixed microscopic property like the bond conductivity or spring constant. The singularities in the macroscopic conductivitivity or elastic modulus of the network are then determined by the cluster geometries, as discussed earlier. We now consider such properties for the 'Swiss cheese' model of continuum percolation, where (say) spherical holes are randomly placed in a uniform medium. Let us consider now such a model where the holes are uniform in size. 'Bonds' are formed whenever the randomly punched spherical holes become neighbouring but do not overlap. See Fig. 1.5(a), where a connected chain of such bonds is formed in two dimensions when random circular holes are punched out from a thin solid. It was shown by Halperin et al. (1985) that the singularities, expressed by the exponents for the conductivity or the elastic modulus of such continuum networks, are considerably different (larger) compared to those for the lattice percolations. This happens mainly because, somewhat similar to the breakdown problem discussed in this book, the transport properties of any such channel formed out of the chains of surviving bonds depend mainly on the transport capacity (determined by the cross-section and length) of the narrowest constriction along the channel. Unlike the breakdown problem, one of course adds up here over the tranport capacities of all such 'parallel' channels, to get the macroscopic transport coefficients. With some reasonable assumptions, one can express the cross-section (and the length) of such 'weakest' bonds along the channel in terms of the percolation cluster statistics on the lattice (in terms of the correlation length ξ in particular). One can therefore express the conductivity $\Sigma \sim (p-p_c)^{\tilde{t}_c}$ and the elasticity modulus $Y \sim (p - p_c)^{\tilde{T}_e}$ for the continuum percolation in the Swiss-cheese model with the conductivity exponent $\tilde{t}_c = t_c + x$ and the elasticity exponent $\tilde{T}_e = T_e + y$, where t_c and T_e represent the conductivity and elasticity exponents of the lattice percolation, with $x = 0$ and $1/2$, and $y = 3/2$ and $5/2$ in $d = 2$ and 3 respectively (Halperin et al. 1985).

We now briefly indicate here how such results are obtained. Figure 1.5(b) shows a typical portion of a channel between two holes, of radius a in two dimensions. If we imagine a rectangular strip of width δ to be formed here, then the length $2l$ of the strip is of the order of $\sqrt{\delta a}$. This is because the length l forms the base of a right-angled triangle with the apex at the centre of a hole and the radius being the adjacent side and the hypotenuse being of length $a + \delta : a^2 + l^2 = (a + \delta)^2$, giving $l \sim \sqrt{\delta a}$. This gives the

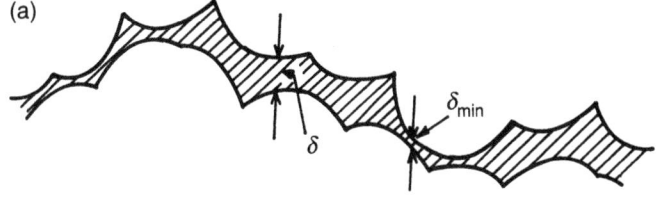

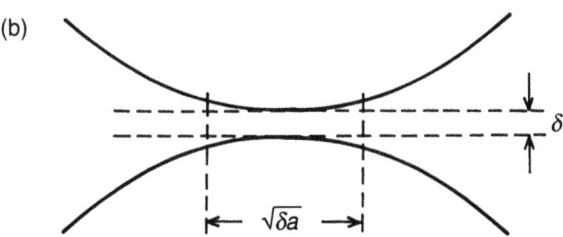

FIG. 1.5. (a) A typical link-element structure in the Swiss-cheese model of continuum percolation in two dimensions. The channel width is denoted by δ. (b) The dashed lines indicate the outline of the rectangular bond which approximates the narrow neck of a channel.

conductivity σ of any such bond with cross-sectional width δ to be given by $\sigma \sim \delta/l \sim (1/\sqrt{a})\delta^{1/2}$ in $d = 2$ and $\sigma \sim \delta^2/l \sim (1/\sqrt{a})\delta^{3/2}$ in $d = 3$. Similarly, one can find the bond-bending force of such strips or bonds. Any channel formed by such randomly punched holes has a distribution $P(\delta)$ of the cross-sectional width δ. It is assumed that $P(\delta)$ is finite as $\delta \to \delta_{min} \simeq 0$. In fact, δ_{min} may be assumed to scale inversely with L_c, the chemical length of a link of the node-link-blob model discussed in Section 1.2.1(d).

As discussed in Section 1.2.1(e), the conductivity Σ of such a model node-link network may be estimated as

$$\Sigma \sim \xi^{-(d-1)}\xi\Sigma_L \sim \xi^{2-d}\Sigma_L, \tag{1.14a}$$

where

$$\Sigma_L^{-1} = \sum_i (1/\sigma_i) \sim \int_{\delta_{min}}^{\infty} P(\delta)\delta^{-x}d\delta; \quad \delta_{min} \sim L_c^{-1}, \tag{1.14b}$$

and Σ_L denotes the conductivity of a link-element. Halperin et al. (1985) assumed $\Sigma_L \sim L_c^{-1}$ in $d = 2$ since the bond conductance varies less rapidly than linearly with δ and hence the sum is not determined by the weakest link; rather all L_c bonds contribute equally to the link conductivity. In $d = 3$, assuming the weakest bond to determine the conductivity of the link and $P(\delta_{\min})$ to be finite, one gets $\Sigma_L \sim \delta_{\min}^{3/2} \sim L_c^{-3/2}$. This gives $\Sigma \sim (\Delta p)^{\tilde{t}_c}$ with $\tilde{t}_c = t_c$ in $d = 2$ and $\nu + 3/2 = t_c + 1/2$ in $d = 3$. Similar calculations for the elastic (rigidity) modulus give $\tilde{T}_e = T_e + y$ for the elasticity exponent in continuum percolation, with $y = 3/2$ and $5/2$ in $d = 2$ and 3 respectively.

1.2.2 Stress concentration and statistics of extremes

(a) Stress concentration

If one applies external force (stress) on an elastic solid containing voids or cracks, or puts an external electric field on a dielectric containing conducting defects, or sends current through a conductor containing dielectric defects, the stress or field or current density in the solid becomes nonuniform. The stress concentrates around the defects and the amount of concentration depends on the geometry of the defects. Since the concept of this stress concentration around defects is very basic to all the breakdown phenomena, we give below a precise analysis and estimate for the stress concentration in a two-dimensional continuum solid containing a single elliptic defect. This was first used by Inglis (1913) for the study of fracture strength of solids. However, being a vector problem, the stress concentration in solids is slightly complicated. We give here the equivalent and simpler derivation (following Duxbury et al. 1987) for the scalar problem of current concentration in a conducting plate containing an elliptic hole or dielectric (see Fig. 1.6).

Let the lengths of the semi-major and -minor axes of the ellipse be $2l$ and $2b$, while the linear size of the conductor is L ($\gg l$ or b). A potential difference $E_0 L$ is applied across the conductor (say, in the y-direction). For obtaining the voltage distribution within the conductor, one has to solve the two-dimensional Laplace equation in the xy plane

$$\nabla^2 V = \frac{\partial^2 V}{\partial^2 x} + \frac{\partial^2 V}{\partial^2 y} = 0, \qquad (1.15)$$

with the boundary condition

$$\left[\frac{\partial V}{\partial \mu}\right]_{\mu=\mu_0} = 0, \qquad (1.15a)$$

along the boundary ($\mu(x,y) = \mu_0$) of the elliptic dielectric, assuming that no current flows across the surface of the conductor to the dielectric. The elliptic boundary of the dielectric is represented by

$$\frac{x^2}{l^2} + \frac{y^2}{b^2} = 1. \qquad (1.16)$$

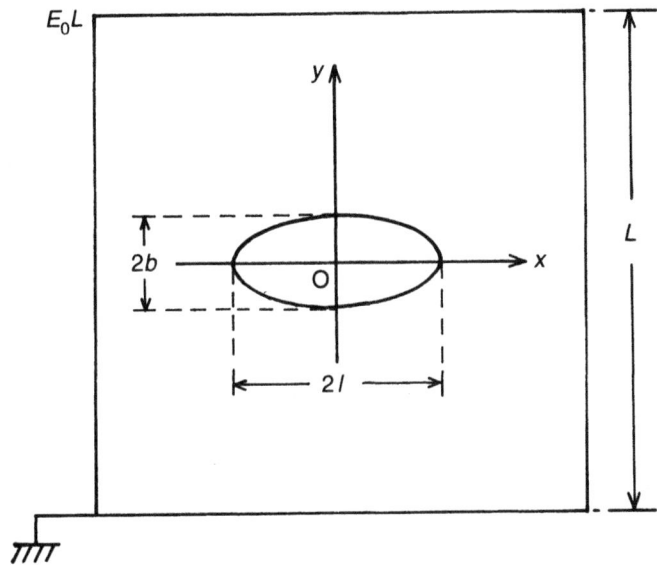

FIG. 1.6. A conducting plate containing an elliptic hole (dielectric) with semi-axes lengths l and b, subjected to potential difference $E_0 L$ at its two ends in the vertical direction. Current density in the plate becomes nonuniform and concentrates at the two horizontal tips of the ellipse.

One can now use the transformations to the elliptic coordinates

$$x = c\cosh(\mu)\cos(\eta); \quad y = c\sinh(\mu)\sin(\eta), \qquad (1.17)$$

with

$$l = c\cosh(\mu_0); \quad b = c\sinh(\mu_0) \quad \text{and} \quad c = \sqrt{l^2 - b^2}, \qquad (1.17a)$$

which satisfies the above equation (1.16) of the ellipse, with $\mu = \mu_0$ (with the tips of the 'crack' at $\eta = 0$ and π). The Laplace equation (1.15) being invariant under the transformations to elliptic coordinates (see Morse and Feshbach 1953), the general solution can be written as

$$V(\mu, \eta) = Ax + A'y + Be^{\mu}[\sin(\eta) + B'\cos(\eta)] + Ce^{-\zeta}[\sin(\eta) + C'\cos(\eta)], \qquad (1.18)$$

where A, A', B, B', C and C' are constants. One can easily check that a straightforward substitution of the above solution in the Laplace equation satisfies it. The boundary condition is satisfied with the choice $A = B = 0 = B' = C'$ and $A' = E_0$, giving

$$V = E_0 y + Ce^{-\mu}\sin(\eta)$$
$$= -cE_0\sinh(\mu)\sin(\eta) + Ce^{-\mu}\sin(\eta), \tag{1.19}$$

with
$$C = -cE_0 e^{\mu_0}\cosh(\mu_0). \tag{1.19a}$$

If Σ denotes the conductance of the conductor, the current density in the y-direction is then
$$i_y = \Sigma \frac{\partial V}{\partial y}$$
$$= -\left(\frac{\Sigma}{c}\right)\frac{\cosh(\mu)\sin(\eta)[\partial V/\partial\mu] + \sinh(\mu)\cos(\eta)[\partial V/\partial\eta]}{\cosh^2(\mu) - \cos^2(\eta)}$$
$$= E_0\Sigma - \left(\frac{C\Sigma}{c}\right)\exp(-\mu)\frac{\cosh(\mu)\sin^2(\eta) - \sinh(\mu)\cos^2(\eta)}{\cosh^2(\mu) - \cos^2(\eta)}. \tag{1.20}$$

One then gets for the current-density concentration at the tip of the ellipse at $\mu = \mu_0$ and $\eta = 0$
$$i_{\text{tip}} = \Sigma E_0 \left[1 + \frac{\cosh(\mu_0)}{\sinh(\mu_0)}\right]$$
$$= \Sigma E_0 \left[1 + \frac{l}{b}\right]$$
$$= i\left[1 + \sqrt{\frac{l}{\rho}}\right], \tag{1.21}$$

where i denotes the current density at large distance away from the elliptic defect and $\rho = b^2/l$ is the radius of curvature at the tip of the ellipse.

The current concentration factor $(1 + l/b) = (1 + \sqrt{l/\rho})$, obtained above in (1.21), is also valid for the stress concentration in a stressed (two-dimensional) solid containing an elliptic void with the semi-major and -minor axes of lengths $2l$ and $2b$, and having curvature ρ $(= b^2/l)$ at the tips of the major axis. One can therefore easily see that if the void is sharp enough ($\rho \to 0$), or if its length $2l$ is very large, the stress concentration can increase several levels above the external stress level (far away from the defect) and the solid may break or fracture (or fuse) from the defect or crack tip.

(b) Extreme statistics

In an earlier section, we have discussed the statistics of clusters of defects, produced for example in the random percolation processes. We have also discussed there how some of the linear responses, like the elastic moduli or

A brief introduction to some theoretical ideas and models 23

conductivity, of such random networks can be obtained from the averages over the statistics of such clusters. This was possible because of the self-averaging property of such linear responses.

As can be guessed from the discussions in part (a) of this section on stress concentration, and discussed later extensively in this book, the breakdown properties of solids are somewhat extreme and non-self-averaging in nature. This is because, unlike say the conductivity of a random network where all the 'parallel' conducting paths contribute their share in the net conductivity of the sample, the fracture or breakdown property of a disordered solid is determined by only the weakest (often the longest) defect cluster or crack in the entire solid. Except for some indirect effects, most of the weak or small defects or cracks in the solid do not determine the breakdown strength of the sample. The fracture or breakdown statistics of a solid sample is therefore determined essentially by the extreme statistics of the most dangerous or weakest (largest) defect cluster or crack within the sample volume. We discuss below the general features of this extreme statistics.

Let us consider a solid of linear size L, containing n cracks within its volume. We assume that each of these cracks has a failure probability $f_i(\sigma)$, $i = 1, 2, ..., n$ to fail or break (independently) under an applied stress σ on the solid, and that the perturbed or stress-released regions of each of these cracks are separate and do not overlap. If we denote the cumulative failure probability of the entire sample, under stress σ, by $F(\sigma)$, then (Ray and Chakrabarti 1985a)

$$1 - F(\sigma) = \prod_{i=1}^{n}(1 - f_i(\sigma)) \simeq \exp[-\sum_i f_i(\sigma)] = \exp[-L^d g(\sigma)], \quad (1.22)$$

where $g(\sigma)$ denotes the density of cracks within the sample volume L^d (coming from the search sum $\sum_i$ over the entire volume), which starts propagating at and above the stress level σ. The above equation comes from the fact that the sample survives if each of the cracks within the sample survives. This is the essential origin of the above extreme statistical nature of the failure probability $F(\sigma)$ of the sample.

As discussed earlier, the percolation theory or similar theories of the cluster statistics in disordered solids can give us the probability density $g(l)$ of the defect clusters of linear size l:

$$g(l) \sim (1-p)^{l^{d'}} \sim \exp[l^{d'}\ln(1-p)], \quad (1.23a)$$

for lattice bond (site) occupation probability $p \to 1$, where the defect dimension is $d' \leq d$, and from the pair correlation function $C(r, p)$ in (1.3), one gets

$$g(l) \sim \exp\left(-\frac{l}{\xi(p)}\right), \qquad (1.23b)$$

for p near the percolation threshold p_c, where ξ denotes the percolation correlation length. It may be noted here that although $g(\sigma)$ and $g(l)$ are different functions, we use the same notation to indicate their common origin.

For use of these cluster statistics $g(l)$ in eqn (1.22) for the failure probability $F(\sigma)$, one needs to relate the breaking stress σ to the length l of a crack, and extract $g(\sigma)$ from $g(l)$. This is usually done using the stress concentration formula (1.21), or some generalised versions of it like the Griffith (1920) formula discussed later in Chapter 3. Let us write

$$\sigma \sim \frac{\Lambda}{l^\psi}, \qquad (1.24)$$

for the stress beyond which a crack of length l breaks. Here Λ is determined by various linear responses (like elasticity, conductivity etc.) of the solid and the exponent ψ is determined by the nature of the stress concentrations etc. Employing (1.23) and (1.24) in the expression (1.22) for the cumulative failure probability, we get the Gumbel distribution (Gumbel 1958, Duxbury et al. 1986, 1987)

$$F(\sigma) \sim 1 - \exp\left[-L^d \exp\left(\frac{\Lambda^{d'/\psi}\ln(1-p)}{\sigma^{d'/\psi}}\right)\right], \quad \text{for } p \to 1, \qquad (1.25a)$$

$$\sim 1 - \exp\left[-L^d \exp\left(-\frac{\Lambda^{1/\psi}}{\xi\sigma^{1/\psi}}\right)\right], \quad \text{for } p \to p_c. \qquad (1.25b)$$

Following the same procedure, a similar double exponential form for the cumulative failure distribution $F(\sigma)$ was obtained by Ray and Chakrabarti (1985a) and was finally reduced (fitted) to a Weibull form (Weibull 1951). Later, Duxbury et al. (1986) obtained this generic form (1.25) using Lifshitz argument for the estimate of the size of the most dangerous defect for small disorder ($p \simeq 1$).

In case of disorder correlations, or for example at the criticality (at $p = p_c$), the probability $g(l)$ of a defect cluster of size l decreases following a power law $g(l) \sim l^{-w}$. Using then the relation (1.24), connecting the failure stress σ with the size l of the crack, one gets $g(\sigma)$, which in turn, when put in (1.22) for the cumulative failure probability $F(\sigma)$, gives the Weibull distribution (Weibull 1951, Ray and Chakrabarti 1985a)

$$F(\sigma) \sim 1 - \exp\left(-\frac{L^d \sigma^m}{\Lambda^m}\right), \qquad (1.26)$$

where the Weibull modulus $m = w/\psi$.

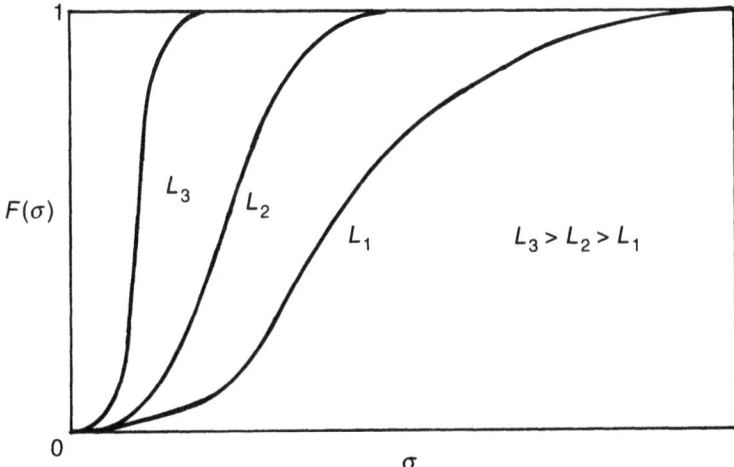

FIG. 1.7. Typical variation of the cumulative failure probability $F(\sigma)$ of a solid sample under stress σ, shown for three large linear sizes L_1, L_2 and L_3.

The cumulative failure probability $F(\sigma)$ of a sample of size L under stress σ has therefore the general feature that $F(\sigma)$ goes to unity as the applied stress increases ($\sigma \to \infty$). As the sample size L increases, this curve generally shifts towards lower values of σ, such that $F(\sigma)$ becomes of the order of unity for even smaller values of σ (see Fig. 1.7). For extremely large L, $F(\sigma) \simeq 1$ for any nonvanishing σ. This feature is true for both the Gumbel and Weibull distributions, and comes from the fact that with increasing sample size the probability of getting an even weaker (or larger) crack increases. Of course, this increase in the probability or the consequent shift in $F(\sigma)$ towards lower values of σ, with increasing system size L, is practically very slow (with extremely small power $1/m$ in the Weibull case and logarithmically slow in the Gumbel case), and is difficult to perceive or observe unless the system size changes by orders of magnitude. It may also be noted at this point that although the failure probability is unity for large stresses ($F(\sigma) = 1$ as $\sigma \to \infty$) for any system size L in the Weibull form (1.26), this obvious feature is not necessarily obtained for the Gumbel distribution (1.25) unless the system size L is very large.

Assuming a finite value (say $\sim 1/2$) for the cumulative failure probability $F(\sigma)$ at the most probable breakdown stress σ_f, one gets from (1.25) and (1.26)

$$\sigma_f \sim \frac{\Lambda[\ln(1-p)]^{\psi/d'}}{(\ln L)^{\psi/d'}}, \quad \text{for } p \to 1, \qquad (1.27a)$$

$$\sigma_f \sim \frac{\Lambda(p)/\xi^\psi}{(\ln L)^\psi}, \quad \text{for } p \to p_c, \qquad (1.27b)$$

from the Gumbel distribution, and

$$\sigma_f \sim \frac{\Lambda(p)}{L^{d/m}}, \qquad (1.27c)$$

form the Weibull distribution. Since Λ and also ξ vary with disorder (Λ being determined by the linear responses of the network), one gets the singularity in the variation of σ_f with p near p_c from the dependence of σ_f on Λ (and also on ξ). However, for any given disorder, both these forms clearly show that the average fracture or breakdown stress vanishes ($\sigma_f \to 0$), although extremely slowly (logarithmically or with an extremely small power), for any disordered solid in the truly macroscopic size limit ($L \to \infty$). This is again due to the extreme nature of the statistics of failure or breakdown; the probability of getting an extremely weak (large) crack existing already within the sample volume, due to statistical fluctuations in the defect concentrations, increases with the volume of the sample.

(c) Extreme statistics versus percolation statistics

As we will see in the appropriate sections of the next two chapters, the precise ranges of the validity of the Weibull or Gumbel distributions for the breakdown strength of disordered solids are not well established yet. However, analysis of the results of detailed experimental and numerical studies of breakdown in disordered solids suggests that the fluctuations of the extreme statistics dominate for the entire range of disorder, even very close to the percolation point.

Bergman has recently argued (see Bergman and Stroud 1992), however, that arbitrarily close to the percolation threshold, the fluctuations of the extreme statistics may get suppressed and the percolation statistics should take over (most probable breaking stress becoming independent of the sample volume, and its variation with disorder being determined by the appropriate breakdown exponent). It has been suggested that the appropriate competing length scales for the two kinds of statistics are the Lifshitz scale $\ln L$ (coming from the finiteness of the volume integral of the defect probability: $L^d (1-p)^l \sim$ finite, giving the typical defect size $l \sim \ln L$) and the percolation correlation length ξ. When $\xi < \ln L$, the above scenario of extreme statistics should be observed. For $\xi > \ln L$, the percolation statistics is expected to dominate. As we will see later when discussing the computer simulations and experimental results for fracture or electrical breakdowns (in the appropriate sections of the next two chapters), the dominance of the percolation statistics in the failure behaviour is indeed very difficult to observe. In fact, it is possible that near the percolation point, the Lifshitz length scale changes to $\xi \ln L$ (as $L^d g(l) \sim$ finite, with $g(l) \sim \exp(-l/\xi)$ near $p = p_c$, giving the typical defect size $l \sim \xi \ln L$). If so, the Lifshitz scale and consequently the extreme statistics will always dominate, even near the percolation threshold, because of the fact that $\xi \ln L$ is always greater than the percolation correlation length ξ.

1.2.3 Self-organised criticality and sandpile models

Many of the complex dynamical dissipative systems in nature, involving many interacting dynamical units, seem to evolve collectively to a self-organised state. A simple example may be a pile of sand which grows on a horizontal table due to random deposition of sand grains. With the sprinkling of the grains of sand on the table, one after another, a pile builds up. The dynamics is such that if the local slope at any point of the surface of the pile grows beyond a critical value, a local avalanche occurs and the sand grains are redistributed. This reduces the local slope anywhere to values less than or equal to the critical value. Eventually, the conical pile ceases to grow, as the additional sand grains ultimately fall off the table through various sizes of avalanches. This self-organised state of the pile, with statistical fluctuations over the average angle of repose, are an attractor of its dynamics. These self-organised states are often critical, in the sense that self-similarities are observed in spatial and temporal structures over all possible scales, leading to fractal-like power law behaviours. It is suggested, and partly observed, that the avalanche sizes in a sandpile, at its self-organised angle of repose, are power law distributed. It appears that many random dynamic systems, undergoing global failure, approach such critical states in a self-organised way, through the stress redistributions due to local failures. We will discuss some such examples in detail later.

In the static critical behaviour of various thermodynamic systems, like in magnets or percolating systems, the average macroscopic quantities follow scaling behaviours, essentially scaled appropriately by the correlation length ξ which is finite away from the critical point (where $\xi \to \infty$), except at the critical point where such scaling behaviours disappear and power law behaviours are observed. In such systems, the power laws occur only at the critical point, which can be achieved by fine tuning of the driving fields, like the temperature or the magnetic field in case of magnets, the dilution concentration in case of percolation, etc. Unlike such systems, the above-mentioned randomly driven dissipative dynamical systems in nature show self-organised criticality, without any fine tuning. Such systems evolve naturally towards the self-organised critical point and stay there self-tuned.

As mentioned before, examples of such systems are truly abundant in nature. The widespread observations of fractal spatial structures in nature, from galaxies to snow-flakes, temporal fractal structures in $1/f$ noise (or $1/f^\phi$ noise with $\phi \sim 1$) in the power spectrums of fluctuating light intensities of quasars, or the current fluctuations in resistors, are all examples of such self-organised critical systems. Here, neither in space, nor in time, has one any particular characteristic scale (length or frequency); such structures are identical in all length and time scales. In fact, the observed earthquake magnitude-frequency dependence, as represented by the Guttenberg-Richter power law (Guttenberg and Richter 1954), also indicates that such dynamic breakdown phenomena are self-similar in all scales, and occur due to their inherent self-organised criticality.

Recently, some very simple cellular automata models of such randomly driven dissipative systems have been developed and have been studied extensively. It has been shown that the dynamics of such models leads to a critical state characterised by power laws induced by stochastically developed self-similarities in the system. One such popular model, known as the BTW model, introduced by Bak et al. (1987, 1988), attempts to capture the avalanche dynamics of a sandpile where the sand grains are being added to the pile at a constant rate. The model has been studied extensively, both numerically and analytically, and the existence of the self-organised criticality in the model has been established.

The BTW sandpile model

Let us consider the BTW model in two dimensions. Let us associate to each lattice site i of a square lattice, an integer variable z_i representing the 'height' of the sandpile at that site. If at any site i, z_i becomes greater than $z_0 = 3$, say, the site 'topples'. On toppling, z_i decreases by 4 (i.e. vanishes there) and the heights z_{nn} of each neighbouring site (of site i) increase by 1. If any of these neighbouring sites had $z_{nn} = 3$, before toppling of the ith site, that will become unstable and will topple subsequently. This may introduce a chain or sequence of topplings, called an 'avalanche', before the system settles to a new stable configuration (where $z_i \leq 3$ for all i). One assumes an absorbing boundary such that topplings at the boundary cause dissipation of the heights (or sand). If one now adds continuously these integer heights z at randomly chosen sites, the above dynamics of the system brings the average height $\bar{z}$ of the sandpile from zero to a critical value z_c, beyond which it does not increase and the additional sands (height) is lost through the boundaries, due to global avalanches. The system self-organises to a stochastic (dynamic) equilibrium state which is critical, characterised by power laws for the probability distributions. Several extensions of the model have been proposed and studied, including a binary version with $z_i = 1$ or 0 (Manna 1991).

A brief introduction to some theoretical ideas and models

The mathematical simplification of the critical height condition (instead of realistic critical slope conditions, etc.) leads to the commutivity of adding heights at two different sites within the bulk of the BTW model of a sandpile. Because of this property, many of the properties of such 'Abelian' sandpile models can be calculated exactly (Dhar 1990). It was shown that the exact number of states in the attractor of the dynamics of the above model is $(3.210...)^{L^2}$, out of the 4^{L^2} possible states in the model of size $L \times L$. Several consequences can then be found for the model analytically. These and other properties, even in some related but different models, have been studied numerically extensively (see e.g. Grassberger and Manna 1990). For example, one finds that at the self-organised critical point, the average height $\bar{z}$ is given by $\bar{z} = z_c \simeq 2.16$ and 2.66 in square and simple cubic lattices. At this critical point, as one adds more height (or sand) to the pile, instabilities are settled through various avalanches. It is found that these marginally stable clusters (which get affected or topple as more sand is added) have self-similar properties, and the (net) avalanche distribution has a power law: $n_s \sim s^{-\tau}$, where n_s denotes the density of s-size avalanches (or s height units coming out of the system at a time) and τ is the avalanche exponent ($\tau \simeq 1.15 \pm 0.10$ in $d = 2$).

As mentioned before, it has been suggested that in many breaking processes within random solids, such self-organisation towards the global avalanches or breakdowns occurs as the local failures induce the redistribution of the stress field among the intact portions (bonds or sites) of the solid, leading to further increase in the failure possibility of the intact portions. In particular, in several earthquake models, the occurrence of such self-organised criticality has been conjectured to be the origin of the observed power law distribution of earthquake magnitudes (see Chapter 4).

2
ELECTRICAL BREAKDOWN IN DISORDERED SOLIDS

2.1 Introduction

In this chapter we shall begin with the problems associated with failures in disordered solids under the influence of an electric field, because this kind of failure is simpler than the mechanical failure or fracture. However, all the cases of failure present some common features, so that the study of electric failures will provide a convenient framework for introducing several important concepts.

We shall consider two extreme kinds of systems. In the first kind, the system is a conductor and by application of a voltage between two electrodes (for the sake of simplicity the two electrodes will be taken parallel) a current flows from one electrode to another. The failure occurs when the current density becomes larger than a threshold value. Consequently, the system becomes nonconducting. The system behaves exactly as a fuse which is destroyed when the current is too large. We shall call this failure the fuse failure. In the second case, the system is a perfect insulator and a voltage is applied between the two electrodes. Again, beyond a definite (threshold) value of the electric field, the system breaks down and becomes conducting. This phenomenon is well-known in the physics of dielectrics, since it limits the application of dielectrics as insulators. We shall talk about the dielectric problem for this kind of failure.

In the present work, we shall not discuss the exact nature of the failure, i.e. its microscopic mechanism. In the fuse problem, the mechanism of the failure is very well-known (it is merely the Joule effect), but in the dielectric problem the mechanism is much more complicated (O'Dwyer 1973). The reason is that we intend to attack the problem from a point of view which is of tremendous importance for statistical analysis. If the sample is perfectly homogeneous the failure will take place in all the portions of the sample. In the first case the current density is uniform in the sample and in the second, the electric field is the same everywhere. If the threshold value is reached, the failure will be general and the sample will explode. In fact, this never happens. The failure always begins as a local event and progressively becomes general. This is because there are weak points in the system. The failure always begins at these weak points. The existence of weak points is due to the fact that solids are never homogeneous. This means that the

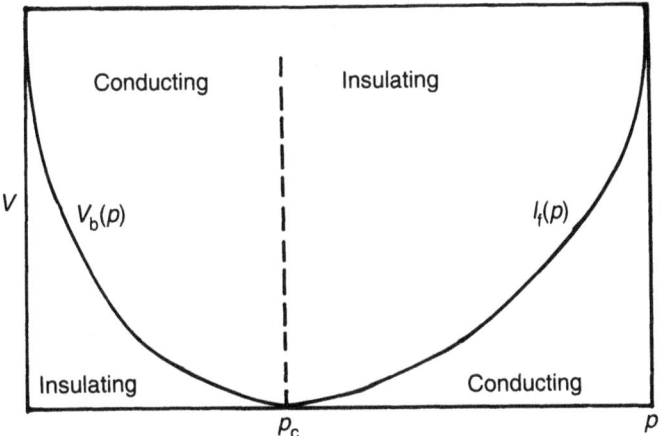

FIG. 2.1. Phase diagram of failure in a mixture of a conductor (with concentration p) and an insulator (with concentration $1 - p$). On the left side, the system is insulating before the failure (electric field V less than the breakdown field V_b) and conducting afterwards. On the right side, the system is conducting before failure (current I less than the fuse current I_f) and becomes insulating afterwards.

current density (fuse problem) or the electric field (dielectric breakdown problem) is not uniform throughout the sample. The weak points are always related to some amount of disorder in the solids.

The problem of the influence of disorder on the physical properties of materials is very important since it has implications in many fields: composite materials, polymers, emulsions, ionosphere, etc. The concept of disorder is very general and in each case it is necessary to state precisely what we mean by disorder.

The first thing we have to specify is the length scale of the disorder. In the present case, since we are not interested in the exact mechanism of the failures, the length scale will be larger than the regions in which the local failure appears. For this reason we shall assume that the defects (consequence of the disorder) are macroscopic. It is difficult to always define an exact length scale but it will be assumed to be much larger than the atomic distances.

The second thing we need to characterise is the type of disorder we shall consider. As mentioned before, a major part of this book will be concerned with the lattice models of disorder, with the statistics governed by that of random percolation. The simplest physical picture of this kind of disorder

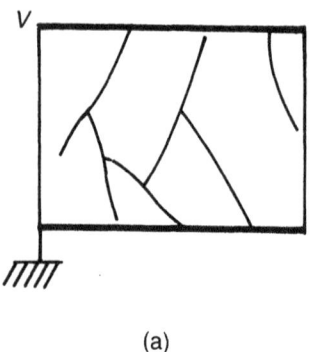
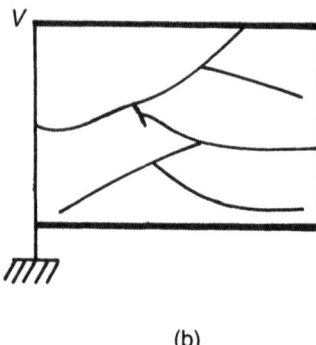

FIG. 2.2. Failure paths. (a) In the dielectric problem, the failure path gives the possibility for a current to flow from one electrode to the other. (b) In the fuse problem, the failure path is made of insulating elements and it prevents the current across the sample.

is to consider random defects made of the 'opposite' material. In the fuse problem the defects are perfectly insulating and in the dielectric problem they are conducting. Hence, the two problems can be seen as concerning the same mixture of two materials: one conducting and the other insulating.

Let the system be made of a perfect insulator and introduce, at random, some conducting inclusions which take the portion p of the volume. Thus for $p < p_c$ (p_c is the percolation threshold) the system is insulating since there is no continuous conducting path between the electrodes. However, for $p > p_c$ the whole sample is conducting.

We can now draw the phase diagram presented in Fig. 2.1. On the left vertical axis, the magnitude of the voltage (field) V applied on a random dielectric is given. The curve $V_b(p)$ indicates the voltage which must be applied to transform the sample into a conductor. On the right side we plot the current flowing through a random fuse system, and the curve $I_f(p)$ indicates the current which must flow in the sample to transform it to an insulator. The main problem is now to determine how V_b and I_f vary with p. As shown in Fig. 2.1, V_b and I_f both go to zero at p_c^- and we shall come back later to explain more carefully this point.

One can ask, what exactly is V_b or I_f? First, we have to imagine how the sample looks after the failure. In the dielectric problem, after the failure has taken place, there is a conducting path composed of conducting portions which connects the two electrodes (Fig. 2.2a). In the fuse problem, after the failure, the current cannot flow since there is now an insulating path more or less perpendicular to the current direction (Fig. 2.2b). It is also possible that other parts of the sample, which do not belong to the 'failure path', have also failed. The important question is: how does the 'failure path' appear?

This is a difficult question which has not yet received a complete answer. But we can discuss qualitatively the failure process. We consider the following experiment. We increase slowly the applied voltage until a local failure takes place. At this point we stop increasing the voltage, and one can have different possibilities. We shall mention only two of them. (i) After the first failure, the whole sample fails by a succession of local failures until complete failure. This is a cascade effect. (ii) After the first failure, nothing happens. One has to increase the voltage for the appearance of new local failures until complete failure.

By analogy with the mechanical fracture (discussed in the next chapter), the case (i) corresponds to a brittle failure and the case (ii) to a ductile failure. Although there is no definite answer, it is believed that in the case of percolation disorder, the electrical failures are mostly brittle-like failures: the voltage (or the current) for the first failure is often the voltage (or the current) for the failure of the whole sample, especially for disorder concentrations near the percolation threshold. We shall see later a different type of disorder which can give ductile failure.

We close this introduction with a final remark about the modelling of the failure. In a real situation, failure takes place in solid samples which are, by nature, continuous in space. However, many studies (numerical and experimental) have been made on lattices. In all these studies, it is an implicit assumption that one can replace a continuous solid by a lattice. For example, a conducting solid can be described by a lattice in which the bonds between sites are identical resistors. It is a very common practice in percolation type models of disordered solids. We stress that this transformation (continuous solid to lattice) defines a particular length scale: the length of the unit cell of the lattice. This implies that defects appear by discrete steps and this does not correspond always to real situations. We shall see later how to remove this limitation.

If this transformation (continuous solid to lattice) does not present a particular problem in electricity, the elastic case is more subtle and needs to be approached with some caution. We shall return later to this point.

2.2 The fuse problem

2.2.1 *Qualitative analysis*

What we need to understand is how the insulating defects within a conductor can modify the current of failure I_f. For that, we shall consider the two limits which can be easily analysed. The first is the dilute limit when the defects are in small quantity, i.e. p is near 1. In such a case, the defects can be seen as isolated units without interaction. The second limit is when

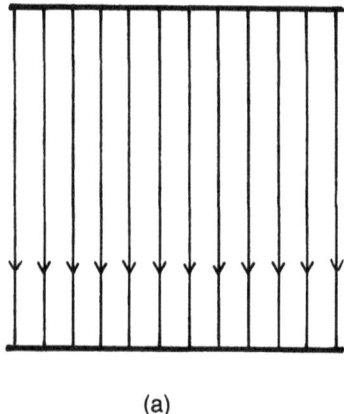 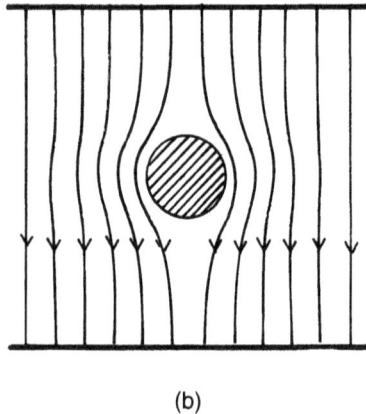

FIG. 2.3. Lines of constant current in a pure sample (a), and in the presence of a defect (b). In (b), there is an enhancement of the current density at the left and right sides of the defect.

the concentration of defects is large, when p is near p_c, and the concept of individual defects has no meaning.

We begin with the system in which there is one defect, with a size equal to the length scale of this particular sample. In a perfect sample, the current lines are all parallel to one another and perpendicular to the electrode surfaces (Fig. 2.3a). In Fig. 2.3(b) we show a sample with one defect which is chosen spherical in three dimensions (3D) or circular in two dimensions (2D). In an intuitive way, one can draw the current lines around the defects when it is supposed that far from the defect the current lines are not perturbed (Fig. 2.3b). The modification of the lines gives an enhancement of the current density immediately to the right and the left of the defect.

One can write this current density i_m as

$$i_m = i(1+k), \qquad (2.1)$$

where k is the enhancement factor which depends on the geometry of the problem (see Section 1.2.2(a)). In (2.1), i is the unperturbed current density far from the defect. The total current I is

$$I = Si = \frac{Si_m}{1+k}, \qquad (2.2)$$

where S is the electrode surface (in 2D it is the electrode length). The first failure occurs when i_m becomes equal to i_0 which is the (threshold) current

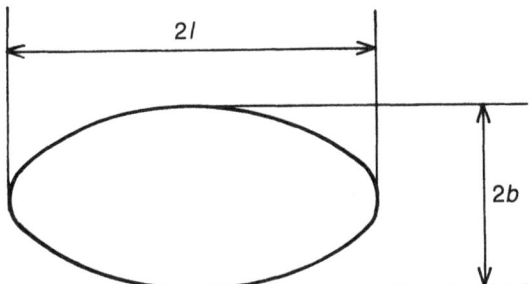

FIG. 2.4. An elliptic defect.

density for the failure of the sample without defect. The failure current is then

$$I_f = \frac{Si_0}{1+k}. \tag{2.3}$$

The enhancement factor for the current density gives a decrease of the failure current I_f. In accordance with what we said in the introduction, the current of the entire first failure here is also the failure current for the sample. Once the regions immediately adjacent to the defect fail, the current density will increase on the sides of this new defect and by propagation the whole sample fails. The sample will be divided into two pieces after the failure.

We thus expect a rapid decrease of the quantity $I_f(p)$ in the vicinity of $p = 1$. We can either have a true discontinuity of the failure current or maybe only an infinite derivative dI_f/dp at $p = 1$. Only an exact calculation can give the answer.

We now consider larger defects but in a small quantity, i.e. we are still in the vicinity of $p = 1$. In this dilute limit there is no interaction between the defects. The important question is: what is the most dangerous defect or, in other words, what is the defect which will introduce the largest enhancement of the current density?

The more dangerous defects will be flat defects perpendicular to the current lines. In two dimensions they are lines, but in three dimensions they have the shape of cylinders. From the results of Section 1.2.2(a), one can see that the enhancement factor for an elliptic defect (see Fig. 2.4) is $k = l/b$. Thus a long defect parallel to the current lines ($b \gg l$) does not affect the current lines, while a long defect perpendicular to the current line ($b \ll l$) affects them strongly.

The next step is to ask what is the probability of occurrence of a defect with a given size. The exact answer will depend on the exact shape permitted for the defects. However, in our case, we need to find the probability of occurrence of the flat defects perpendicular to the current lines. At this stage only a statistical analysis will get the correct answer. As we shall see in the following section, one can calculate the size of the most probable defect and the important result is that not only does it depend on $(1-p)$ (which is the volume fraction of the insulating defects) but also on the ratio of the sample size to the length scale of the problem. A large sample will give a large most probable defect. A more precise analysis on a lattice model gives that $I_f \sim 1/(\ln L)$ in two dimensions or $I_f \sim 1/(\ln L)^{1/2}$ in three dimensions, when L is the linear size of the sample. A crucial property of the fuse current I_f is that it decreases when L increases. For an infinite sample, I_f goes to zero.

Near p_c, we have additional problems. The current now flows through the infinite cluster made of all the possible paths from one electrode to the other. This infinite cluster can be seen as very tortuous arms connected at the nodes (see Section 1.2.1(d)). The current will be concentrated in a small number of links such that for $p \sim p_c$, the last link may be broken by a very small current in the sample. This means that I_f goes to zero when p goes to p_c. Near p_c, the failure current is related to the correlation length ξ by

$$I_f \sim i_0 \frac{a}{\xi} \quad (2D), \qquad I_f \sim i_0 \frac{a^2}{\xi^2} \quad (3D), \tag{2.4}$$

where a is the thickness of a link and the number of such links is proportional to ξ^{-1} and ξ^{-2} in dimensions $d = 2$ and 3 respectively. Since the correlation length diverges at p_c with the exponent ν, one gets that near p_c

$$I_f \sim (p - p_c)^{\nu} \quad (2D), \qquad I_f \sim (p - p_c)^{2\nu} \quad (3D). \tag{2.5}$$

We recall the values of the correlation length exponent ν: they are 1.33 in $d = 2$ and 0.9 in $d = 3$. The dependence of I_f on the size L of the system, as discussed earlier, is not clearly understood or settled here for $p \to p_c$ (see Sections 1.2.2(c) and 2.1.2(c)).

2.2.2 Quantitative analysis: most probable failure current and distibution

(a) Lattice percolation

A quantitative analysis of the failure process was made by Duxbury et al. (1987) by modelling the system by a lattice and we shall present their results. The simplest lattices were taken: a square lattice in two dimensions and a simple cubic lattice in three dimensions, in which the bonds are all equal resistors to begin with. Each resistor can stand a current up to i_0. If $i > i_0$, the resistor is fused and becomes a perfect insulator. It is believed that the results are not dependent on the type of the lattice as it was proven in the case of percolation. The size of the lattice is L: in two dimensions it is a square and in three dimensions a cube with L unit cells per side. We now consider the effect of a fraction $(1-p)$ of randomly removed bonds (insulators) in the network. As above, we shall consider the two limits: $p \to 1$ and $p \to p_c$.

Dilute limit: the most probable failure current   The smallest defect which can be made is to remove one resistor. If it is one of the resistors perpendicular to the current flow (horizontal resistors), nothing happens since in the perfect lattice only resistors parallel to the current flow carry current. The simplest defect is the absence of one of the parallel resistors (vertical resistors) in the sample. For the time being, we suppose that the new defect is far from the boundaries of the sample. The problem of defects near the boundaries was investigated by Li and Duxbury (1987).

The enhancement coefficient was calculated in two dimensions, and it is equal to $4/\pi$. This means that for $p \to 1$, I_f is decreased by the factor $\pi/4$. For a perfect sample in $d=2$

$$I_f = Li_0, \qquad (2.6)$$

and for the sample with one defect

$$I_f = \frac{\pi}{4} Li_0. \qquad (2.7)$$

We shall now determine what is the probability of occurrence of a dangerous defect as defined in the preceding section. In the present case of a lattice, it consists in n neighbouring resistors belonging to the same plane perpendicular to the current flow. In $d=2$, it is merely a line of n removed bonds and in $d=3$, it is an ensemble of n removed bonds forming a hole with approximately the shape of a disc. The current in the immediately adjacent parallel resistor will be in accordance with the enhancement factor in an elliptic defect

$$i_m = i(1 + k_2 n) \quad (2D), \qquad i_m = i(1 + k_3 n^{1/2}) \quad (3D). \qquad (2.8)$$

In three dimensions, we have to recall that the enhancement factor is related to $n^{1/2}$ because the current which is diverted by n defects is spread uniformly around the perimeter of the defect. And the perimeter is proportional to $n^{1/2}$. In (2.8), i is the current flowing through one vertical resistor located far from the defect.

The probability that n vertical resistors have been removed to form a dangerous defect is

$$P(n) \sim (1-p)^n L^2 \quad (2D), \qquad P(n) \sim (1-p)^n L^3 \quad (3D), \qquad (2.9)$$

where $(1-p)^n$ is the probability of n resistors missing and the L^d term comes from the 'volume', giving the number of places that the defect can occupy. When this probability is of the order of unity, we have the most probable dangerous defect and if we denote by n_c the value of this defect size, we get

$$n_c = -\frac{2}{\ln(1-p)} \ln L \quad (2D),$$

$$n_c = -\frac{3}{\ln(1-p)} \ln L \quad (3D). \tag{2.10}$$

From (2.8) and (2.10), one deduces the current i_m in the resistor adjacent to the defect:

$$i_m = i\left[1 + k_2\left(\frac{-2\ln L}{\ln(1-p)}\right)\right] \quad (2D),$$

$$i_m = i\left[1 + k_3\left(\frac{-3\ln L}{\ln(1-p)}\right)^{1/2}\right] \quad (3D). \tag{2.11}$$

Since the total current in the sample is $iL^{(d-1)}$, one gets the total failure current by putting i_m equal to i_0, the threshold value of the current through the sample:

$$I_f = \frac{i_0 L}{1 + K_2[\frac{\ln L}{|\ln(1-p)|}]} \quad (2D),$$

$$I_f = \frac{i_0 L^2}{1 + K_3[\frac{\ln L}{|\ln(1-p)|}]^{1/2}} \quad (3D). \tag{2.12}$$

The constants K_2 and K_3 are equal respectively to $2k_2$ and $\sqrt{3}k_3$. As expected, I_f depends on both p and L. We shall examine these dependences separately.

First, we note that if p goes to 1, I_b reduces to the value of the perfect lattice. Secondly, it is easy to see that the slope of the curve $I_b(p)$ at $p = 1$ is infinite, in accordance with what we said in the preceding section.

However, the most remarkable result of this study is that the failure current is dependent on the size of the sample. The failure current per bond $i_f = I_f/L^{(d-1)}$ decreases with $\ln L$. If L is large enough and p not very near to 1 (such that the absolute value of $\ln(1-p)$ is not too large) i_f is proportional to $1/\ln L$ and $1/(\ln L)^{1/2}$ in two and three dimensions respectively. This result has direct implications in practice if one recalls that L is the ratio between the real size of the lattice and the length of the unit cell. We recall also that the size of the smallest defect is one unit cell. If in a solid sample, the defects are made of elementary defects, the smaller the size of this elementary defect, the smaller the failure current (but clearly to a certain limit). One can also formulate this result differently: for a given size of the elementary defect, larger and larger samples give smaller and smaller failure currents.

Dilute limit: distribution of failure currents Until now, the analysis concerns the most probable size of the dangerous defect or the most probable failure current. We intend to discuss now the distribution of the failure current, for which the general form was discussed in Section 1.2.2(b). Duxbury et al. (1987) calculated the distribution of the failure currents, using a scaling approach for determining the function $C(n)$ which gives the probability that no defect of size greater than n occurs in a lattice of (linear) size L. This lattice is divided into small cubes (in three dimensions) or small squares (in two dimensions) of linear size L_1. Because of the statistical independence of the small cubes (or squares), the probability that no defect of size greater than n occurs is $[C(n)]^N$. Here N is the number of the L_1 cubes. In order to ensure that the distribution functions have the same form on the L and the L_1 lattices, one requires that

$$[C(n)]^N = C(a_N n + b_N), \qquad (2.13)$$

where a_N and b_N are scaling functions that tend to finite limiting values for N going to infinity. Two general solutions can be found: one if $a_N = 0$ and the other if $b_N = 0$. In the first case, $C(n)$ is given by

$$C(n) = \exp[-b\exp(-cn)] \quad (b, c > 0) \qquad (2.14)$$

and in the second

$$C(n) = \exp(-r/n^m) \quad (r, m > 0). \qquad (2.15)$$

It is easy to verify that the expressions (2.14) and (2.15) represent the effective solutions by inserting them in eqn (2.13). The expression (2.14) must go to zero for $n = 0$ and this is possible if b is large enough. Duxbury et al. (1987) gave several arguments in favour of (2.14). Here we choose a more intuitive method to select the convenient solution.

The probability to find a defect of size n is given by the first derivative of $C(n)$ and the maximum value of this probability must be obtained when n is equal to n_c given by (2.10). Comparing the value of n for which the second derivative of $C(n)$ is equal to zero and n_c, one finds very simple relations between the constants b and c appearing in (2.14) and L and p. However, using (2.15) one has very complicated expressions for the constants r and m. Thus, the simplest expressions which are in accordance with the exact results of Duxbury et al. are chosen. From the zero of the second derivative of $C(n)$ one finds that $n_c = (\ln b)/c$ and consequently from (2.10)

$$b \sim L^2 \quad (2D), \qquad b \sim L^3 \quad (3D), \qquad (2.16)$$

and

$$c \sim -\ln(1-p). \qquad (2.17)$$

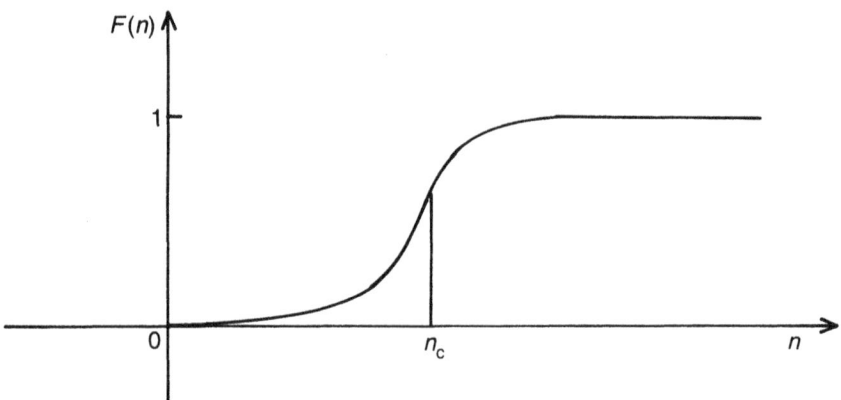

FIG. 2.5. The function $F(n)$ gives the probability of occurrence of a defect with size smaller than n. The maximum of the derivative appears for n_c, which is the most probable size for a defect.

As expected, b is large to ensure that $C(0) \approx 0$. The general shape of the curve $F(n)$ is given in Fig. 2.5 with an inflexion point at $n = n_c$. In particular, one can calculate its slope at $n = n_c$ and it is found that it is proportional to $-\ln(1-p)$. This shows that the distribution is sharp for p very near to 1 and becomes broader when p decreases.

We can write now the distribution of the failure currents by using the expressions (2.8), (2.12), (2.14), (2.16) and (2.17). Thus the cumulative probability of failure for a given current I will be given by (see also Section 1.2.1(b); using same notations $F(n)$ and $F(I)$ to indicate common origin)

$$F(I) = 1 - \exp\left[-AL^2 \exp\left(-2A \ln L \frac{I_0/I - 1}{I_0/I_f - 1}\right)\right] \quad (2D), \qquad (2.18a)$$

$$F(I) = 1 - \exp\left[-AL^3 \exp\left(-3A \ln L \left(\frac{I_0/I - 1}{I_0/I_f - 1}\right)^2\right)\right] \quad (3D), \quad (2.18b)$$

where A is a constant, which is different in two or three dimensions. These expressions are different from those given by Duxbury et al. (1987). We prefer them because the failure current of the pure sample I_0 and the most probable failure current I_f appear explicitly. We recall here that I_f is dependent on both p and L (see eqn (2.12)). This distribution probability is also different from (but equivalent to) that frequently used in the analysis of reliability: the Weibull distribution (see Section 1.2.2(b))

$$F(I) = 1 - \exp\left[-r\left(\frac{I}{I_f}\right)^m\right]. \qquad (2.19)$$

From these cumulative failure distribution probabilities $F(I)$, one gets the probability $P(I) = dF/dI$ that the failure current takes a given value I, and the most probable failure current is that at which $P(I)$ is maximum. It is easy to see that I_f, as it appears in (2.18), is not the most probable value, at variance with what is claimed when I_f is calculated above. However, by a simple but tedious calculation, it is possible to show that, if L is large enough, I_f is practically the most probable value as defined by the maximum of $P(I)$.

The expressions (2.18) for $F(I)$ are plagued by another flaw, which does not invalidate them provided that the sample size is large enough. In principle, $F(I)$ is defined for values of I extending from zero to infinity when F goes from zero to unity. However, it is clear from (2.18) that I is limited such that $I \leq I_0$. For $I = I_0$, $F(I_0) = 1 - \exp(-AL^d) \sim 1$ if L is large enough. Specifically, for example in d dimensions, one has $F(\infty) = 1 - \exp[-AL^d \exp(-dA \ln L/(I_0/I_f - 1)^{d-1})]$, which tends to unity only for large enough L.

Regarding the distribution (2.19), two remarks can be made. First, it is not possible to get it straightforwardly from the second choice for the function $C(n)$ given by (2.15). Secondly, the constant I_f is practically equal to the most probable failure current if the Weibull parameter is large enough (for example, if m is larger than 5).

To conclude, one can say that the function $F(I)$ in (2.18) has the properties of a probability distribution only if the sample has a large enough size. Unfortunately, it is not possible to give a criterion to know what is a 'large sample' since the constant A appearing in (2.18) is unknown.

Failure current for p near p_c We saw in the preceding section that the failure current decreases when p approaches p_c and becomes zero at the threshold. We shall propose a different determination of the failure current from the node-link-blob picture of the infinite cluster (see Section 1.2.1(d)). Instead of using the current I (current flowing through the sample), we shall consider the tension V between the two electrodes. The relation between I and V is given by $V = RI$ where the resistance of the sample is dependent on p,

$$R \sim (p - p_c)^{-t_c}, \qquad (2.20)$$

where t_c is the conductivity exponent: $t_c = 1.33$ and 2 in two and three dimensions respectively.

The infinite cluster is seen as a super-lattice with the unit cell size equal to the correlation length. The mean current in a link is given by $i_L = V_L/R_L$, where V_L and R_L are respectively the voltage accross a link and the resistance of the link. Since there are L/ξ number of links in series along the length of the sample across which the external voltage V is applied, one gets $V_L \sim \xi V$. From the discussions in Section 1.2.1(e), one has $R \sim R_L$

in $d = 2$, and $R \sim \xi R_L$ in $d = 3$, giving $i_L \sim \xi V/R$ and $i_L \sim \xi^2 V/R$ in two and three dimensions respectively. Since V equals the failure or fuse voltage V_f when the current in a link reaches the threshold value which a singly connected resistor in a link can stand, one gets

$$V_f \sim (p - p_c)^{-t_f}, \qquad (2.21)$$

with $t_f = t_c - \nu$ in $d = 2$ and $t_f = t_c - 2\nu$ in $d = 3$. From the relation between V_f and I_f ($I_f = V_f/R$) and taking into account (2.21), one finds again the expression (2.5) giving the dependence of I_f on p.

Since in two dimensions t_c and ν are of the same order of magnitude, one finds that V_f goes to a finite value at p_c. However, in three dimensions, the exponent t_f is about 0.2 and V_f diverges for $p \to p_c$ contrarily to the failure current, which always goes to zero at p_c.

(b) Continuum percolation

In continuum percolation (see Section 1.2.1(g)), we suppose that the defects are introduced in a solid sample as randomly placed insulating holes with the shape of a circle (in two dimensions) or a sphere (in three dimensions) and we include the possibility of overlap of the defects (Swiss cheese model). This last possibility gives near p_c an infinite cluster with the the links having different cross-sectional width δ. This property is essentially responsible for the differences between lattice and continuum percolations.

In the very dilute limit $p \to 1$, there is clearly no difference between the two types of percolation. We thus expect, as above, a strong variation of $I_f(p)$ near $p = 1$. In the dilute limit, when the defects are well separated but their number is not very small, the same argument of dangerous defect that we used in the case of lattice percolation can again be applied. Now L is the ratio between the size of the sample and the diameter of one defect. All the results we established above are valid here.

We expect a difference near p_c. Since there are $(L/\xi)^{(d-1)}$ number of parallel links between the two electrodes, the current in a link is given by $i_L = (\xi/L)^{(d-1)} I$. The current density i in a channel of cross-section δ is given by

$$i \sim \frac{\xi I}{\delta} \quad (2D), \qquad i \sim \frac{\xi^2 I}{\delta^2} \quad (3D). \qquad (2.22)$$

The largest current density is through channels with minimum cross-section $\delta_{\min}$, which is proportional to $1/L_c$, the length of the shortest chemical path (see Section 1.2.1(g)). Thus from (2.22), one sees that if i_0 is the maximum current density that the material can stand, then

$$i_0 \sim \xi L_c I_f \quad (2D), \qquad i_0 \sim \xi^2 L_c^2 I_f \quad (3D),$$

or,

$$I_f \sim (p - p_c)^{\nu+1} \quad (2D), \qquad I_f \sim (p - p_c)^{2(\nu+1)} \quad (3D). \tag{2.23}$$

The fuse voltage is thus $V_f \sim |p - p_c|^{\tilde{t}_f}$, with the exponent $\tilde{t}_f$ equal to $\nu + 1 - \tilde{t}_c$ in $d = 2$ and $2(\nu + 1) - \tilde{t}_c$ in $d = 3$. Here $\tilde{t}_c$ is the conductivity exponent in the continuum percolation: $\tilde{t}_c \simeq 1.3$ and 2.5 in $d = 2$ and 3 respectively (see Section 1.2.1(g)). One thus gets $\tilde{t}_f \simeq 1$ and 1.3 in two and three dimensions respectively.

The principal results are that the exponents of the failure current are higher than those of discrete percolation and that the failure voltages have a completely different behaviour. While in the present case of continuum percolation, they always go to zero at p_c, for lattice percolation the failure voltage either reaches a finite value (in two dimensions) or even diverges (in three dimensions). This just reminds us that the physical quantity which brings the failure is the local current density.

For the fuse problems with various other criteria of failure, for example breakdown due to local Joule heating in a random thermal fuse model, see Sornette (1987) and Sornette and Vanneste (1992) and also Section 2.2.6.

(c) Influence of sample size and distribution probability

We shall now investigate the influence of the sample size. As discussed earlier, it is related to the notion of the most dangerous defect in the sample. In the present case, the most dangerous defect is a 'cell' of the infinite cluster with length ξ in the direction parallel to the current and l_{max} in the perpendicular direction. The total probability of getting a defect of size l is given (using $g(l)$ from eqn. 1.23b) by

$$g(l) \left(\frac{L}{\xi}\right)^d \sim \exp\left(-\frac{l}{\xi}\right) \left(\frac{L}{\xi}\right)^d, \tag{2.24}$$

and l_{max} is obtained when this probability is of order unity:

$$l_{max} \sim \xi \ln L. \tag{2.25}$$

Now, the current flowing through the side link of this 'defect' is proportional to $l_{max}I$ in $d = 2$ and $(l_{max})^2 I$ in $d = 3$. Using a similar argument as given above, one finds that $I_f \sim i_0 l_{max}^{-(d-1)}$, or

$$I_f \sim \frac{(p - p_c)^{(d-1)\nu}}{(\ln L/\xi)^{(d-1)}}. \tag{2.26}$$

As Bergman has pointed out (Bergman and Stroud 1992), and discussed earlier (in Section 1.2.2(c)), this behaviour is expected when the Lifshitz scale is greater than the percolation correlation length. It is not clear however if the Lifshitz scale is given by $l_{max} \sim \xi \ln L$ or simply by $\ln L$ near p_c.

Of course, in case it is given by $l_{\max}$ ($\gg \xi$), the extreme statistics will always dominate over percolation statistics. We will discuss the experimental observations later.

To find the probability distribution, we use the results of Section 1.2.2(b). We recall that near p_c, it is found that the general form of the cumulative failure distribution $F(I)$ is (using $\Lambda^{d'} \sim (p-p_c)^{(d-1)\nu}$ in eqn. 1.25a)

$$F(I) = 1 - \exp\left[-L^d \exp\left(-\frac{K(p-p_c)^{(d-1)\nu/\psi}}{I^{1/\psi}}\right)\right], \quad (2.27)$$

where the exponent ψ is equal to 1 in $d=2$ (since $I_f \sim l^{-1}$) and 2 in $d=3$ (since $I_f \sim l^{-2}$). This distribution is very similar to the double exponential distribution that we found for very low p. So far, there is no experimental confirmation of this probability distribution.

(d) Some discussion on quantitative analysis

We give here a brief résumé of the theoretical results of the failure behaviour in the fuse model.

It was found that the failure current I_f depends not only on the fuse (resistor) concentration p, but also on the size of the sample through $\ln L$. The dependence of I_f on $\ln L$ was discussed more fully by Li and Duxbury (1987). They also investigated the role of different dangerous defects in this sense, that in their vicinity there is a possibility of strong enhancement of the current. They proposed to define an exponent ψ such that I_f depends on L through $(\ln L)^{-\psi}$, with an approximate bound for this new exponent:

$$\frac{1}{2(d-1)} < \psi < 1, \quad (2.28)$$

where d is the dimension of the sample.

It was proposed by Duxbury and Li (1990) to condense the results of the preceding analysis in an interpolation formula like

$$I_f = I_0 \frac{\left[\frac{(p-p_c)}{(1-p_c)}\right]^\phi}{1 + K\left[-\frac{\ln(L/\xi)}{\ln(1-p)}\right]^\psi}. \quad (2.29)$$

In this expression (2.29), the different exponents and the constant K depend on the dimension and on the type of percolation. The expression (2.29) has been checked in the following limits :
(i) For $p=1$ we find $I_f = I_0$.
(ii) Near $p=1$, $(p-p_c)$ is almost constant and we have again the expressions (2.12).

Table 2.1 *Theoretical estimates for the fuse exponent t_f*

dimension (d)	Lattice percolation	Continuum percolation
2	ν 1.33	$\nu + 1$ 2.33
3	2ν 1.8	$2(\nu + 1)$ 3.8

(iii) Near p_c, we recover the behaviour of I_f near p_c as proportional to $(p - p_c)^\phi$. The values of the exponent ϕ ($= t_f$ here) are given in Table 2.1. From (2.26) and (2.12), one can see however that an unique formula is possible only in two dimensions.

2.2.3 Numerical simulations and experimental results

(a) Numerical simulations

The first numerical simulations of the fuse problem were made by those who proposed the model for the first time (de Arcangelis *et al.* 1985). The simulation consists in calculating the currents in a two-dimensional resistor lattice where $(1-p)$ bonds are removed when an external voltage is applied. The voltage is chosen larger and larger until the current in one bond is equal to the failure current of one resistor. It is the failure voltage for the first failure. One can also calculate the current in the whole lattice to get the current for the first failure. As mentioned above, it is also the current for a complete failure of the sample. However, the propagation effect which brings the complete failure from one local failure is very complicated since it is now a dynamic problem. To our knowledge, there is no theoretical description of the cascade effect which can give answer to different questions such as what is the failure path or what is the number of broken bonds.

De Arcangelis *et al.* proposed the following procedure. Once the current in one bond reaches the failure value, the voltage is suppressed and the fused bond is removed. The voltage is applied again and one looks for the bond with the maximum current. The voltage for this new local failure is determined, the voltage is suppressed, the bond is removed and the above procedure is repeated until the whole sample becomes non-conducting. In general, the successive values of these failure voltages decrease during this process. De Arcangelis *et al.* determined the voltage for which the first bond is fused (and we can take this value for the failure voltage V_f) and the voltage V_f^{fin} necessary to fuse the last bond. For square lattice, the behaviour they found for V_f and V_f^{fin} are very different when p decreases from unity: V_f first decreases until $p \sim 0.7$, where it reaches a minimum and then increases again; V_f^{fin} increases regularly and becomes very near

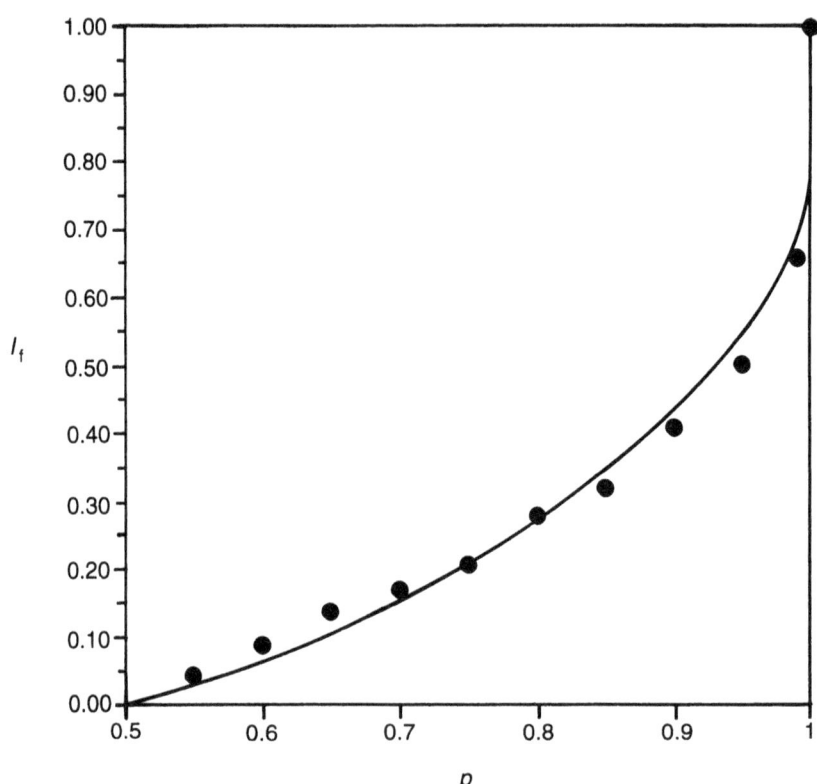

FIG. 2.6. Fit of the computer simulation results for the fuse current I_f (from Duxbury et al 1987) with the expression (2.29). We took $\ln(L/\xi)$ constant since $L \gg \xi$.

to V_f for $|p - p_c| < 0.08$. These increases of V_f^{fin} and of V_f near p_c were interpreted as a divergence with an exponent 0.48. This is not in agreement with the theoretical result: in two dimensions it was found that V_f goes to a finite value for $p = p_c$. It seems that in this dimension V_f exhibits only a pseudodivergence and this is indicated in the observations presented by de Arcangelis et al. (1985).

Duxbury et al. (1987) performed the same type of simulations as that of de Arcangelis et al. but they covered the whole range of p from 1 to $p_c = 0.5$. They also checked the dependence of I_f on L and the distribution of the failure currents. These simulations were also made on a square lattice.

Their calculation of the dependence of I_f on p gives the possibility to check the interpolation formula (2.29) that was proposed above, assuming

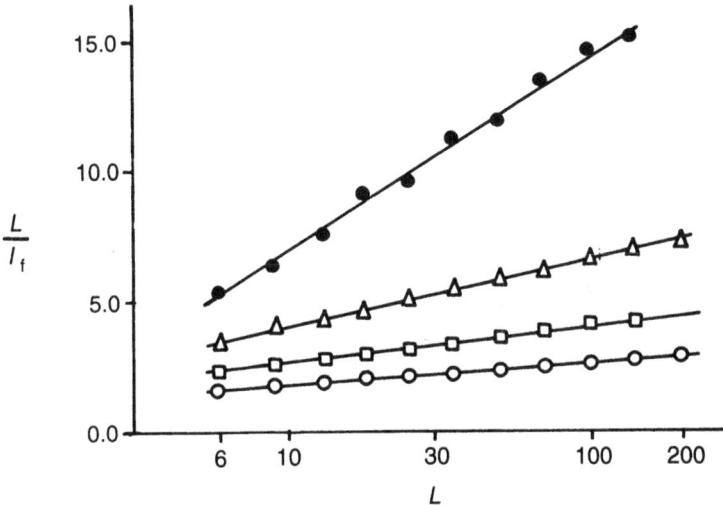

FIG. 2.7. L/I_f versus $\ln L$ showing their linear dependence (from Duxbury et al. 1987). The curves from top to bottom correspond to $p = 0.6, 0.7, 0.8$ and 0.9 respectively.

$L \gg \xi$. In Fig. 2.6, we show the results of the simulation from the results of Duxbury et al. and a fit with (2.29). One can see good agreement. However, to get a satisfactory fit, it is necessary to take $\phi = 1$, instead of 1.33, as expected from the above theoretical analysis. This means that it is better to take the exponent ϕ in (2.29) as an effective one.

Duxbury et al. determined also the dependence of V_f on p and their result is very similar to that of de Arcangelis et al. The difference is only in the interpretation. The finite value of V_f at p_c is demonstrated by comparing the variation of I_f and of the conductance near p_c.

Duxbury et al. (1987) determined by numerical simulations the dependence of I_f on L, varying L from 10 to 200. For fixed value of p, the I_f values are determined as functions of L. Following (2.12), L/I_f must vary linearly with $\ln L$, since the simulations were performed in two dimensions. Their results are presented in Fig. 2.7, and one can see a good agreement with (2.12). From this agreement, one can deduce that the exponent ψ in (2.29) is equal to 1. The slope of the straight lines giving L/I_f versus $\ln L$ increases when p decreases towards the percolation threshold.

The distribution probability $F(I)$ was also calculated and compared to the double exponential or Gumbel distribution, and to the Weibull distribution. The expression (2.18) was not compared directly, but Duxbury et al. (1987) preferred the equivalent expression

$$F\left(\frac{V_f}{L}\right) = 1 - \exp\left(-AL^2 \exp\left(-\frac{KL}{V_f}\right)\right). \tag{2.30}$$

If the calculated $F(V_f)$ is plotted in regular coordinates, it is not possible to distinguish between (2.30) and (2.19). However, by plotting $\ln[\ln(1 - F(V_f))]$ versus L/V_f for the Gumbel distribution and also versus $\ln(V_f)$ for the Weibull distribution, it is possible to show that the most appropriate expresssion is that given by (2.30), the Gumbel or double exponential distribution.

Finally, we mention that de Arcangelis et al. (1985) as well as Duxbury et al. (1987) have determined the number N_f of fused or broken bonds up to complete failure. Both groups found that N_f decreases to 0 for $p \to p_c$ with an exponent near that of the correlation length.

(b) Experimental results

There are very few experimental results which were planned to verify the theoretical results presented above. We mention the work of Gilabert et al. (1987) on a two-dimensional lattice of resistors. They verified the behaviour of I_f for very small values of $1 - p$ and found that effectively I_f depends on $1-p$ as in expression (2.12). They tried also to determine the behaviour of I_f near p_c but they got an exponent smaller than the expected one ($\nu = 1.33$) because the size of the sample was too small.

In conclusion to this section we can say that the numerical simulations and the experiments are in full agreement with the theoretical expectations. They were made only in two dimensions and it is clearly desirable to have more experimental results, particularly in three dimensions.

2.2.4 Other kinds of disorder: distribution of the failure threshold

Until now, we supposed that the disorder was of the random percolation kind. We describe here another kind of disorder in the case of the resistor lattice. Only the two-dimensional case has been studied. The disorder comes from the fact that the resistors do not have the same failure current or failure voltage. Although in the fuse problem the current is the relevant quantity, Kahng et al. (1988) developed this model considering only a distribution of threshold voltage.

A simple model is to suppose that the failure voltage for a resistor is distributed uniformly between two limits v_- and v_+, where

$$v_- = 1 - w/2,$$

$$v_+ = 1 + w/2. \tag{2.31}$$

The case $w = 0$ corresponds to a sample without disorder and the case $w = 2$ to a uniform distribution between 0 and 2. The model is studied using the following method. A voltage is applied to a sample of size L, until one bond fails. The voltage is kept constant and this bond removed. The voltage in the remaining resistors is calculated: if there is another bond which fails, it is removed and a new search for another failed bond is made. This process goes on until the whole sample is failed: the failure is brittle. This is not exactly the cascade effect that we mentioned at the beginning of this chapter. The procedure here is a succession of static problems whereas the cascade effect is a dynamic one. Now, after removing the first fused resistor, under the same external voltage, if there is no other fused bond, the voltage is increased. The process is repeated until the sample is broken: the failure is ductile. Kahng et al. (1988) suggest that it is a good model for 'slow breaking': after the failure of one bond, another bond fails only after the currents in the system have time to take their equilibrium values.

The interesting point of this investigation is that the failure is ductile or brittle depending on w and L. There exists a value w_0 of w for which whatever the value of L, the failure is always brittle. This is the trivial regime. For $w > w_0$, the failure is brittle for large values of L, and ductile for small L. The separation between the two regimes is given by a function $w_c(L)$ such that for $w \to 2$ and $L \to \infty$ the regime is brittle.

To understand the failure in this model, an approximate criterion for the brittle or ductile fracture was proposed. One considers the sequence of the weakest bonds and asks for the average failure voltage for the nth weakest bond. It is given by

$$v_1 = \langle V_{\text{weak}}(n) \rangle = v_- + \frac{nw}{L^2}. \tag{2.32}$$

In (2.32), the average failure voltage is linear in n since the distribution is uniform and must be equal to v_- for $n = 0$ and to v_+ for $n = L^2$. Suppose now that n bonds have failed, forming $2n$ edge bonds where there is an increase of the current due to the enhancement effect (see Section 2.2.1). The average failure voltage for these $2n$ bonds is found to be

$$v_2 = \langle V_{\text{edge}}(n) \rangle = v_- + \frac{w}{2n+1}. \tag{2.33}$$

v_2 is thus a decreasing function of n, since by increasing n the probability to include weak bonds in the $2n$ edge bonds is increased. An approximate criterion for a brittle situation or the instability of the system is

$$kv_1 > v_2, \tag{2.34}$$

where k is the enhancement factor for a single defect. We saw above that for a square lattice $k = 4/\pi$ (see Section 2.1.1). By drawing the curves $kv_1(n)$ and $v_2(n)$ (see Fig. 2.8) one can see that there are two possibilities.

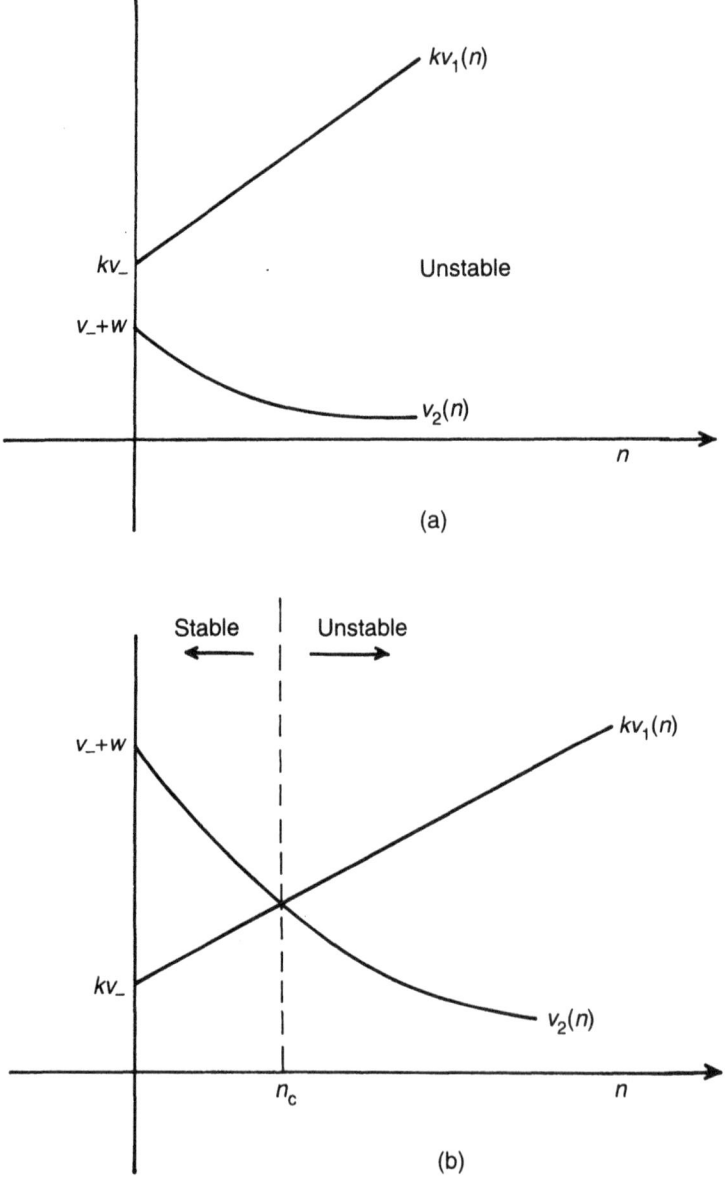

FIG. 2.8. The condition $kv_1 > v_2$ indicates that the fuse system is unstable, once the first resistance is fused; here if one keeps the voltage constant, the successive failures will make the failure total (brittle failure). If however $kv_1 < v_2$, one has to apply a larger voltage after the first breaking until the failure is total (ductile failure).

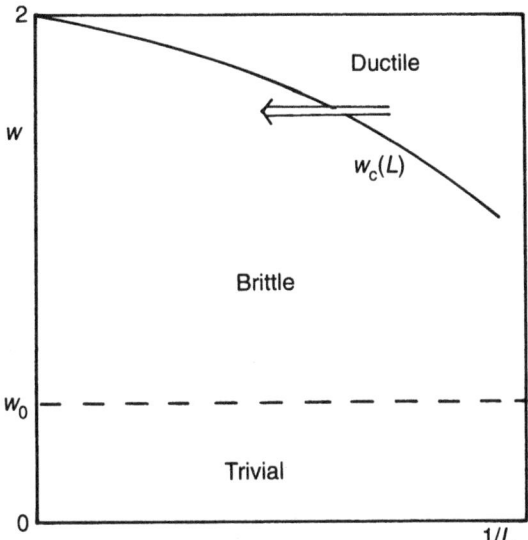

FIG. 2.9. The phase diagram of the failure for the model of distributed thresholds (from Kahng et al. 1988).

In the first, there is no crossing of curves and the system becomes unstable after the first failure, whatever the value of L. This occurs if kv_- is larger than $v_- + w$. Noting that $v_- = 1 - w/2$, one gets the value of w_0 equal to $2(k-1)/(k+1) \simeq 0.24$, below which the randomness becomes trivial. This is because the minimum voltage required for the first weakest bond to break is $v_- = 1 - w/2$. This breaking creates a voltage concentration kv_- at its edges. However, the next breaking due to this voltage concentration is certain only if $kv_- \geq v_+ = 1 + w/2$. The equality gives the above estimate for w_0.

In the second possibility a crossing occurs for n_c. Thus the system is stable during the appearance of the n_c first defects and after it becomes unstable and fails. However, if the sample is small, it can happen that the system is never unstable and fuses in a ductile manner. This argument shows the possibility of two types of failure. The final results (from Kahng et al. 1988) are given in Fig. 2.9.

The variation of the mean failure voltage (per bond) v_f was also studied and different behaviour was found in the brittle and ductile regimes. In the case of brittle failure, it is found that

$$v_f = v_- + H\frac{w}{L^2}, \qquad (2.35)$$

but in the ductile regime,

$$v_\text{f} \sim (\ln L)^{-\psi}, \qquad (2.36)$$

where H is some constant and the exponent ψ is about 0.8. This last result received confirmation by Leath and Duxbury (1994) who studied the fuse model with a continuous threshold distribution starting from zero (ductile regime). Since v_f cannot be smaller than v_- it is easy to understand from (2.35) that by increasing L the system goes into the brittle regime.

To close this section, we mention that this case of disorder was extended to other kinds of threshold distributions, like the Weibull distribution (de Arcangelis 1990).

2.2.5 The shortest path and the electromigration fuse model

We introduce in this section the notion of the shortest path in a resistor lattice, which is a pure geometrical notion. We shall see that some cases of failure can be described by this geometrical picture. In this section we present also a variant of the fuse model, the electromigration model. The interest of the notion of shortest path is that it can be used in the case of the dielectric breakdown. In the following part of this chapter (Section 2.3 on dielectric breakdown) we shall describe a dielectric model and its experimental realisation which are well analysed with the help of the shortest path approach.

We consider a random resistor square lattice in which a fraction $1 - p$ of the resistors are removed. We define a path on this lattice as follows: a walker begins to walk from one side of the lattice (for the requirements of the electromigration model we consider the lateral sides, i.e. those without electrodes). The walker jumps from one cell to the neighbouring cell by crossing the lattice bonds, irrespective of whether the bond is occupied by resistor (or fuse) or absent. The walk is a self-avoiding walk and after visiting a number of bonds, the walker reaches the opposite side. Only those walks which cross the lattice from one side to the other are considered. The path is composed of all the bonds that the walker visited, and let there be n_0 resistors and n_1 insulating bonds along the path. The path with the minimum value of n_0 is the shortest path. An example of the shortest path and a longer path is given in Fig. 2.10. This path is defined for a given configuration of missing resistors. The mean shortest path is the average over a large number of configurations. The connection with the fuse problem is clear if one supposes that the walker can break each resistor when crossing it. When the walker has crossed the whole sample, there is a complete failure. The shortest path is the path which corresponds to the smallest number of resistors to fuse. We saw above that the breakdown process is related to the resistors with the largest current (hottest bonds) and not with the shortest path. Nevertheless this notion has its own interest.

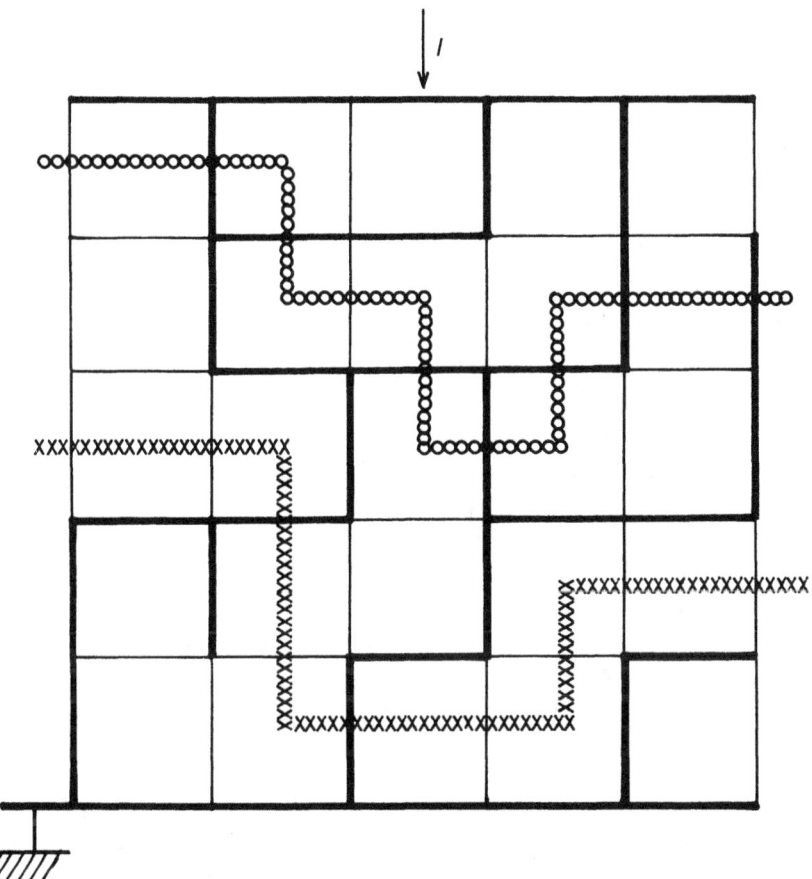

FIG. 2.10. Shortest path (× × ×) and a longer path (ooo) for a walker on a square lattice. The walker jumps from one cell centre to the neighbouring one by crossing the bonds. Here the thick bonds represent the resistors (or fuses) and the thin bonds represent the insulators. In this example the shortest path includes only two occupied bonds (resistors) and the longer path includes seven.

The properties of the mean shortest path has been studied theoretically near the percolation threshold (Chayes *et al.* 1986, Stinchcombe *et al.* 1986) and numerically in the whole range of p (Duxbury *et al.*, unpublished). The important property is the variation of $g = \langle n_0 \rangle / L$, the mean number of resistances in the shortest path, with p. For p near 1, g decreases from 1 linearly with a slope dg/dp about 3 (for the square lattice). In the vicinity of p_c, g goes to zero as $(p - p_c)^\nu$ where ν is the correlation length exponent.

In thin metal films, if the current density becomes very high, collisions between the conduction electrons and the metal ions often lead to slow drift of the ions. This process, known as electromigration, is now well studied and the ionic diffusion, induced by the 'electron wind', is also well understood (Huntington 1975). In polycrystalline thin films, the electromigration of ions leads to random voids or hillocks in the system. As discussed earlier, current concentrations occur at the edges of such voids, inducing further migration and further growth of the voids, until conduction in the circuit ceases and the failure of the circuit is complete. This is a very common feature in miniaturised integrated circuits. It appears that such electrical failures occur dynamically and collectively due to the growth and overlap of various voids in the circuit, with the passage of charge or time. Several models have been developed recently, and we will discuss here one such fuse model (Bradley and Wu 1994, Wu and Bradley 1994).

The electromigration fuse model of Bradley and Wu (1994) is a modification of the fuse model discussed above. As in the fuse model, the electromigration model was studied on a square lattice of resistors. The criterion for a resistor to break is not some value of the current (as in the fuse model) but the total charge that crossed this resistor from the time of application of the voltage. For a constant current I_0 the whole sample breaks after a time τ. The problem is to determine τ for given value of p. For a single bond, the time of failure t_1 is given by

$$\int_0^{t_1} I(t)\, dt = Q_0, \qquad (2.37)$$

where Q_0 is the critical or threshold charge which, when passed through a resistor, makes the resistor fail. For a pure sample τ is given by LQ_0/I_0 and τ decreases with p. For $p \to p_c$, τ goes to zero, since the number of bonds to be broken goes also to zero.

For an isolated defect of length n, it is possible to show that the bond located at a distance n_1 from the edge of the initial defect breaks at time $t_1 = n_1 t_0$ when $t_0 = Q_0/I_0$. Thus the time τ is given by

$$\tau = (L - n)Q_0/I_0. \qquad (2.38)$$

If p is very near to 1, the mean time $\langle \tau \rangle$ is related to the longest probable defect with size n_c. As in the fuse model, n_c is given by (2.10). This means that the bonds which fuse are also the bonds with largest current.

However, this situation changes when the number of removed resistors increases. Not only the bonds lateral to the largest defect break but also the edge bonds of other defects. As the usage time of the (thin film) circuit increases, the charge accumulated in the edge bonds of almost all the existing defects becomes large enough such that the lateral bonds fuse one after the other. This analysis permits one to have some insight into the breakdown path in the electromigration fuse model and to compare it with the breakdown path in the fuse model. In the electromigration model, there are a large number of defects (from the numerical simulation of Wu and Bradley (1994) almost all the defects) which increase their size, but this is not the case in the fuse model. From the numerical simulation of Duxbury et al. (1987) and of de Arcangelis et al. (1985), it is clear that in the fuse model almost all the fused bonds belong to the breakdown path.

From a quantitative point of view, it was argued by Wu and Bradley (1994) that the mean failure time $\langle \tau \rangle$ is given by

$$\langle \tau \rangle = gt_0; \quad g = \langle n_0 \rangle / L, \tag{2.39}$$

and indeed it is a very good approximation. From their numerical estimation, it appears that the relative error is less than 0.7%. One can give a crude argument to understand how a kinetic problem is replaced by a geometrical one. Since almost all the defects grow laterally (perpendicular to the mean current flow) it appears that the coalescence of two defects appears only if they are in adjacent rows and in this manner the final breakdown path is very near to the shortest path.

Equation (2.39) is good for almost all the values of p (except very near $p = 1$, as mentioned above), in particular near p_c. Thus, $\langle \tau \rangle$ goes to zero as the failure current as

$$\langle \tau \rangle \sim (p - p_c)^\nu. \tag{2.40}$$

We have four quantities which go to zero at p_c with the same exponent: the shortest path, the failure current I_f and the number of broken bonds N_f in the fuse model and the failure time $\langle \tau \rangle$ in the electromigration fuse model, and one can ask if this similarity has a deeper reason. Although there is only numerical evidence concerning the exponent of N_f, we admit that it is effectively the correlation length exponent. Concerning $\langle \tau \rangle$, its proportionality with g becomes better and better when approaching p_c.

Consider the largest clusters made by the removed bonds (insulating clusters). Their mean size is the correlation length ξ. The broken bonds at failure are the bonds which connect the largest clusters and their number goes as ξ^{-1}. We can have the same arguments for the bonds belonging to the shortest path and this gives a basis for the claim that $N_f = gL$. Above (see Section 2.2.1) we argued that the failure current in two dimensions is also proportional to the inverse of the correlation length. But this time it is the correlation length of the infinite cluster of the resistors (conducting clusters). However, the exponent of the correlation length for the infinite conducting cluster is the same as the exponent of the correlation length of the insulating clusters and this explains why I_f and N_f have the same behaviour near p_c. This is the basic argument concerning the equality of the exponent for the four quantities mentioned above (g, I_f, N_f and $\langle \tau \rangle$) in two dimensions.

2.2.6 Effects of temperature, AC fields and nonlinearity

Until now, in dealing with the problem of electrical failure, we made two assumptions. First, we always consider a conductor element to have a linear relationship between voltage and current. Secondly, we suppose that there is no influence of the temperature variations occurring due to local Joule heatings. In the fuse problem, the physical process being the Joule heating effect, a conducting element gets heated prior to breaking. The failure arises when some elements reach their melting temperature.

(a) Theoretical analysis

(i) We begin with the case of steady currents (DC) and we adopt a criterion for the occurrence of failure different from that considered earlier. To be specific, we shall consider only the lattice percolation models and we shall not introduce the influence of the sample size. The new criterion is as follows: a conducting bond will be transformed into an insulator if the temperature of the bond reaches a specific value T_m which corresponds to melting.

In a first approach, we take again the picture of the infinite cluster as a super-lattice. If i_L is the current in a link, the change ΔT in the temperature of the bonds in the link will be proportional to $i_L^2 R_L$, where R_L is the link resistance. In $d = 2$, we have $i_L = (\xi/L)I$ and $R \sim R_L$, where I is the total current through the sample. One thus gets

$$I_f \sim (p - p_c)^{\nu + t_c/2}. \tag{2.41}$$

In $d = 3$, we have $i_L = (\xi/L)^2 I$ and $R \sim \xi R_L$, and one gets

$$I_f \sim (p - p_c)^{3\nu/2 + t_c/2}. \tag{2.42}$$

In a second approach, the picture of a link between two nodes of an infinite cluster will be made more precise by introducing the blobs. In a link, the bonds with the largest current (in fact the total current flowing through the link) are the singly connected (sc) bonds. The new melting criterion is applied only to these bonds, by writing

$$\Delta T_{sc} \sim r i_L^2, \tag{2.43}$$

where r is the resistance of a bond. In (2.43) we neglect the change of the resistance due to the heating. Assuming that the failure occurs when ΔT_{sc}

reaches the critical value and using the relations between the total current and the current in a link, one recovers that

$$I_f \sim (p-p_c)^\nu \quad (2D), \qquad I_f \sim (p-p_c)^{2\nu} \quad (3D). \tag{2.44}$$

In conclusion, the two criteria, namely that the current in a link or that the temperature of a singly-connected bond reaches a specific value at the failure, give the same results. It was in fact expected, since the two quantities ΔT_{sc} and i_L are related by the relation (2.43). We shall see below the experimental results giving support for the second approach and making the fuse model a realistic one.

(ii) We will consider now the fuse behaviour of the same percolating random resistor networks under an applied AC voltage across the sample. In fact, under AC fields, new phenomena occurs with the appearance of the third harmonic and we will show that the fuse failure is related to that.

Suppose a current I flows through a sample of resistance R_0. R_0 is defined as the resistance of the sample at zero current. Because of the Joule effect, the resistance will increase by $\Delta R = R_0 \beta \Delta T$, where β is the temperature coefficient of the resistance. As above, one can write $\Delta T \sim R_0 I^2$, and this approximation therefore gives $\Delta R \sim (R_0 I)^2$. The voltage V across the sample is therefore given by

$$V = R_0 I + K R_0^2 I^3,$$

where K is a coefficient of proportionality. If the current varies with time t as $I = I_0 \cos(\omega t)$, then the voltage V across the sample is given by

$$V = R_0 I_0 \cos(\omega t) + V_{3f} \cos(3\omega t), \tag{2.45}$$

where

$$V_{3f} = \frac{1}{2} \Delta R I_0. \tag{2.46}$$

The voltage is thus a sum of two parts: one with the same frequency as that of the applied current and the other with a frequency three times the original frequency – the so called third harmonic. The third harmonic coefficient B is defined as $B = V_{3f}/(I_0)^3$.

We suppose now that our sample is a random lattice resistor network with $1-p$ fraction of missing resistors, where $p > p_c$ such that the lattice is still conducting. We shall now show that there is a relation between the failure current I_f of such a network and its third harmonic coefficient B (Yagil et al. 1993).

Because of the Joule effect, each individual resistance r will increase by an amount δr_j, depending on the current i_j flowing through it. Equating the power generated by the increased resistance of the whole sample with the sum total of the power generated by all the individual increased resistances, we get

$$\Delta R(I_0)^2 = \sum_j \delta r_j(i_j)^2. \tag{2.47}$$

From (2.46), (2.47) and the definition of B, one gets

$$B \sim \frac{\sum_j \delta r_j(i_j)^2}{(I_0)^4}, \tag{2.48}$$

or since $\delta r_j \sim i_j^2 r^2$,

$$B \sim \frac{r^2 \sum_j (i_j)^4}{(I_0)^4}. \tag{2.49}$$

In eqn (2.49), the currents that will give the main contribution are those through the singly-connected bonds, and with this assumption B becomes

$$B \sim \frac{r^2 N_{sc}(i_{sc})^4}{(I_0)^4}, \tag{2.50}$$

where N_{sc} is the number of singly-connected bonds in the sample. The increase in the temperature of the singly-connected bonds (which we suppose equal for all such bonds) is given by

$$\delta T_{sc} \sim r i_{sc}^2. \tag{2.51}$$

Taking into account that $N_{sc} = L_c/\xi^d$, $i_{sc} = (\xi/L)^{(d-1)} I_0$ and using $L_c \sim |p - p_c|^{-1}$, one gets

$$B \sim \frac{|p - p_c|^{-1} \delta T_{sc}}{(I_0)^2} \quad (2D), \qquad B \sim \frac{|p - p_c|^{-1} \xi^2 \delta T_{sc}}{(I_0)^2} \quad (3D). \tag{2.52}$$

These expressions help us to understand why B diverges at p_c. Besides the terms $|p - p_c|^{-1}$ and ξ, δT_{sc} also diverges at p_c. This is because, as one approaches p_c, the number of links decreases and therefore the current through a link increases, giving rise to the increase in the temperature in the singly-connected bonds. We now define a new exponent for the variations of B:

$$B \sim R^{2+w} \sim |p - p_c|^{-t_c(2+w)}. \tag{2.53}$$

Equating now the total current I_0 in (2.52) with the failure current I_f, when δT_{sc} reaches a threshold value, one has

$$I_f \sim B^{-x}$$

with

$$x = \frac{1}{2} - \frac{1}{2t_c(2+w)} \quad (2D), \qquad x = \frac{1}{2} - \frac{1+\nu}{2t_c(2+w)} \quad (3D). \qquad (2.54)$$

In fact, this is a lower bound for the exponent x since we included all the singly-connected bonds in a link. An upper bound is obtained if one takes $N_{sc}\xi^{(d-1)}$ equal to one, when only one of the singly-connected bonds melts, and one gets $x = 1/2$.

To find the exponent w, it suffices to recall that I_f goes to zero with the exponent $(d-1)\nu$. Using this in (2.54), one gets $w = [2(\nu - t_c) + 1]/t_c \simeq 1$ in $d = 2$ and $w = [5\nu - 2t_c + 1]/t_c \simeq 0.73$ in $d = 3$, in agreement with the results of Wright et al. (1986). Using this upper bound for w, we get $0.5 > x > 0.36$ in $d = 2$ and $0.5 > x > 0.32$ in $d = 3$.

We close this section with a remark about the third harmonic coefficient. It is also related to the $1/f$ noise of such networks. However, we shall not develop this relationship here, as it is a little bit away from our main subject (see Rammal et al. 1985a,b, Yagil and Deutscher 1992).

(b) Experimental results

Yagil et al. (1992, 1993) performed a series of measurements of the properties of thin film networks of silver and gold, for which they measured resistance, third harmonic coefficient, failure current and the nonlinear conductivity as functions of the surface coverage by the films. They found that effectively their samples behave in accordance with the random percolation model we discussed earlier. The gold samples have a slightly complicated behaviour, and we will discuss here the results for silver films. The failure current is defined as the current giving the first irreversible change in the resistance of the film. However, contrary to what we said about the fuse model, there is no cascade effect. The resistance decreases for the failure current. This is interpreted as the fusion of the hot points, without disconnection but rather increasing the contact surface between them.

For these films, it was found that $t_c \simeq 1.45$ (Yagil 1992) and the failure current decreases with an exponent value near 2.6, which suggests a continuum Swiss cheese model. However, the value for the exponent w, as determined by them, is around 1.2, which is slightly above the value in lattice percolation (about 1.1) but much smaller than that expected in the Swiss cheese model (in the range 2.7 to 4.2; Garfunkel and Weissman 1985). On the other hand, the exponent value for x relating I_f and B is around 0.48, in good agreement with the upper bound for this exponent (see Fig. 2.11).

Very recently an interesting experiment on the failure under the action of currents has been performed in samples made of epoxy resin loaded

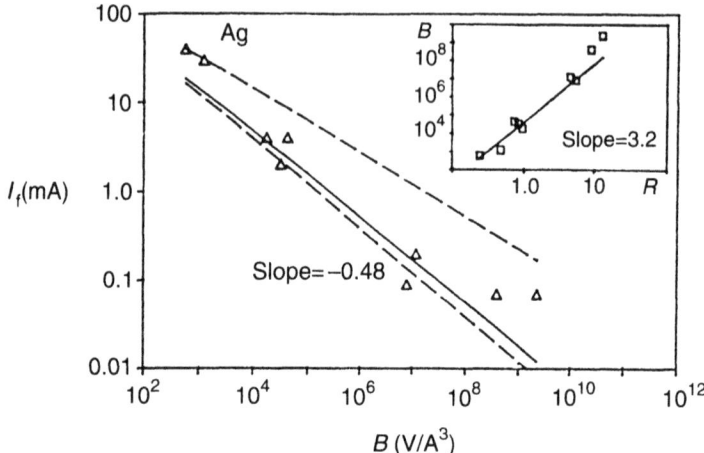

FIG. 2.11. Failure current I_f versus the third harmonic coefficient B showing that the exponent $x \simeq 0.48$ ($I_f \sim B^{-x}$). The inset shows that $B \sim R^{2+w}$ with $w \simeq 1.2$ (from Yagil et al. 1992).

with small spherical carbon particles (Lamaignere et al. 1996). The particular interest in these results comes from the fact that they are dynamic measurements.

The samples were percolating with the conductor concentration above the percolation threshold ($p \simeq 0.34$, when $p_c \simeq 0.23$). A current I, larger than the failure current I_f, is applied and the resistance R of the sample starts increasing with time. In principle, R must go to infinity after a fuse time t_r. From the results of Lamaignere et al., it appears that R diverges following a power law as $t \to t_r$. In fact, they did not observe the divergence because of the limited power the generator used in the experiment could deliver. Nevertheless, the results were fitted to a power law $R \sim (t_r - t)^{-\alpha}$, and the value of α was found to be around 0.65. From the fit of the data for $R(t)$, they also determined the functional dependence of the fuse time t_r on the current I, and they found that $t_r \sim I^{-2}$, for large values of I.

At the present state, only some qualitative explanations are available. When the sample is prepared, the carbon particles get compressed to come in contact with others. During the flow of the current, the particles are heated because of the Joule effect and the heat is transmitted to the surrounding epoxy matrix. As a result of the matrix dilatation, the contacts between the carbon particles disappear and the sample becomes an insulator. The process being gradual, one observes the regular increase of R with time.

2.3 The dielectric breakdown problem

2.3.1 *Qualitative analysis*

As in the fuse problem, we begin with the explanation of how the presence of defects can increase the local breakdown field. A defect is a local change in the properties of the sample. In an insulator, the defects are conducting parts of the sample. We consider again a spherical defect (circular in two dimensions) and we draw the equipotential surfaces or lines (in two dimensions). In a pure sample, these surfaces or lines are parallel to the electrodes (Fig. 2.12a) but in a sample with one defect they show distortions near it. For a two-dimensional sample, the new equipotential lines are shown in Fig. 2.12(b). One sees that in the vicinity of the defect there is an increase of the field. The sample will break at an applied voltage smaller than the one which is needed to break a pure sample. This is the enhancement effect identical to that of the fuse problem and consequently the curve $V_b(p)$ will exhibit an infinite slope when p goes to zero.

We shall now ask what are the most dangerous defects in the case of the dilute limit ($p \to 0$). Whatever the dimension of the sample, they are always very long defects. This is the main difference from the fuse problem. To break the sample, it suffices that a long defect increases by a succession of local breakdowns until a linear or quasilinear conducting path is made. We shall repeat the same kind of argument as in the fuse problem to find the breakdown voltage but only for long and linear defects. It is also the physical reason for the fact that, near p_c, the breakdown voltage goes to zero with the correlation length exponent, irrespective of the dimension.

2.3.2 *Duality in two dimensions*

The dielectric breakdown problem can be solved very easily from the solution of the fuse problem in two dimensions using the concept of duality. This concept is largely used in the case of composite materials and in percolation for problems in $d = 2$ or with cylindrical symmetry (Mendelson 1975, Bowman and Stroud 1989). Here we follow the derivation of Bowman and Stroud.

One considers a composite material made of a mixture of conductors and insulators. As above, for $p < p_c$, the material is insulating, and conducting for $p > p_c$. We consider first the case with p smaller than p_c. The equations for the induction vector D and the field E are:

$$\nabla \cdot \boldsymbol{D} = 0$$
$$\nabla \times \boldsymbol{E} = 0 \qquad (2.55)$$
$$\boldsymbol{D}(r) = \epsilon(r)\boldsymbol{E}(r),$$

where ϵ is the dielectric constant of the insulator. The properties of the material, such as ϵ, depend on the position r. E and D therefore depend

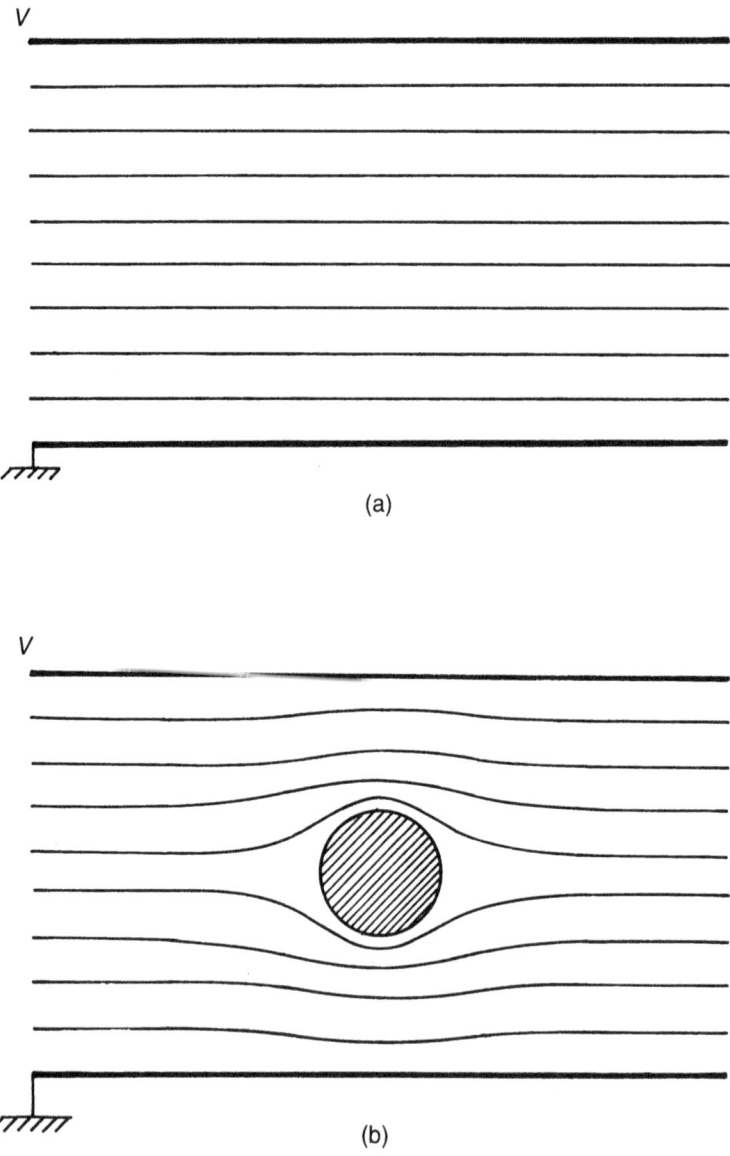

FIG. 2.12. Equipotential lines for a pure insulating sample (a) and for a sample with one conducting defect (b). Note the field enhancement above and below the defect.

on r. The last eqn (2.55) is defined only on the insulating parts of the material. E is also equal to the negative gradient of the scalar potential ϕ, so from (2.55)
$$\nabla \cdot (\epsilon \nabla \phi) = 0, \tag{2.56}$$
or
$$\frac{\partial}{\partial x}\left[\epsilon(r)\frac{\partial \phi}{\partial x}\right] + \frac{\partial}{\partial y}\left[\epsilon(r)\frac{\partial \phi}{\partial y}\right] = 0. \tag{2.57}$$

We consider now the dual composite in which the conducting parts are replaced by the insulating material and vice versa. We suppose that the conductivity of the parts which were insulating is proportional to the inverse of the dielectric constant of the insulating parts ($\epsilon = 1/\sigma$). The material is a conductor and the relevant quantities are the current density i and the field $\bar{E}$. The equations for these quantities are

$$\nabla \cdot \boldsymbol{i} = 0$$
$$\nabla \times \boldsymbol{E} = 0 \tag{2.58}$$
$$\boldsymbol{i}(r) = \sigma(r)\boldsymbol{E}(r).$$

The last eqn (2.58) is valid on the conducting parts, i.e. the insulating parts of the original problem.

Since the divergence of i is zero, it may be expressed as the curl of a potential vector V. We choose it such that only the component V_z perpendicular to the plane of the sample is different from zero, $V_z = \psi(x, y)$. V satisfies the equation
$$\nabla \times \frac{1}{\sigma}(\nabla \times \boldsymbol{V}) = 0. \tag{2.59}$$

Using eqn (2.58), it is found that the scalar potential obeys
$$\frac{\partial}{\partial x}\left[\frac{1}{\sigma(r)}\frac{\partial \psi}{\partial x}\right] + \frac{\partial}{\partial y}\left[\frac{1}{\sigma(r)}\frac{\partial \psi}{\partial y}\right] = 0. \tag{2.60}$$
or
$$\frac{\partial}{\partial x}\left[\epsilon(r)\frac{\partial \psi}{\partial x}\right] + \frac{\partial}{\partial y}\left[\epsilon(r)\frac{\partial \psi}{\partial y}\right] = 0. \tag{2.61}$$

Comparing (2.61) and (2.57), one has
$$\frac{\partial \psi}{\partial x} = \frac{\partial \phi}{\partial x}, \qquad \frac{\partial \psi}{\partial y} = \frac{\partial \phi}{\partial y}. \tag{2.62}$$

The current density i has components
$$i_x = \partial \psi/\partial y = E_y, \qquad i_y = -\partial \psi/\partial x = -E_x. \tag{2.63}$$

Thus we see that the current density is equal in magnitude to the field E, but is rotated by 90° from the dielectric problem. It is also easy to see that

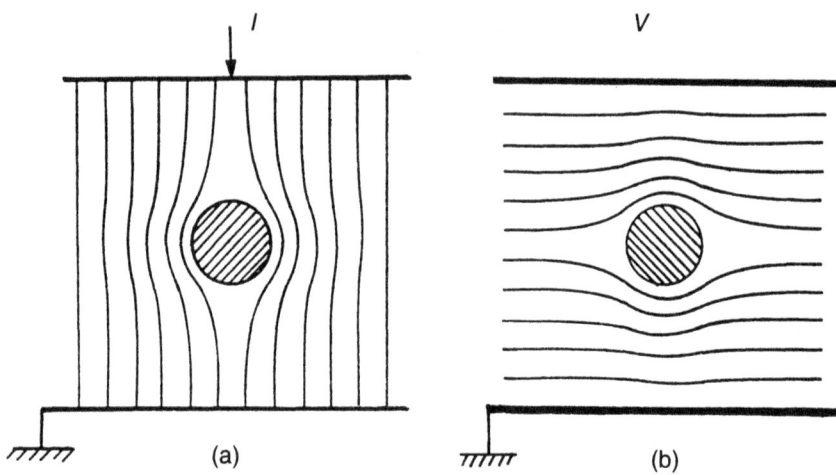

FIG. 2.13. Comparison of constant current lines in the fuse problem and the equipotential lines in the dielectric problem in two dimensions. The two figures are identical but rotated by 90°.

the field $\bar{E}$ of the dual problem is equal to vector D but also rotated by 90°.

We can visualise the equivalence of the two problems by comparing Fig. 2.3(b) and Fig. 2.12(b), which we draw again in Fig. 2.13. We see that the current is zero inside a defect (fuse problem) while the field is nil inside a defect (dielectric problem). Also, we see that the regions with enhancement of the field are located perpendicularly to those with enhancement of the current density.

All the results we got in studying the fuse problem can thus be utilised for the present case. The formula (2.29) can be used when one replaces I_0 and I_f by V_0 and V_b and also $1 - p$ by p, as with formula (2.27).

2.3.3 Lattice percolation

(a) The dilute limit

We shall use the same machinery as that developed above in the case of the lattice fuse model. In order to mimic the insulator part, one takes capacitors with a given value C_1 as insulating bonds. For the conducting bonds, one takes very large capacitor C_0 such that $C_0 \gg C_1$ (Beale and Duxbury 1988). By this choice the voltage across a C_0 capacitor is zero, as within the conductor the field must be zero. p is the fraction of C_0 capacitors which are now seen as defects. Of course, the other possibility is to take resistors for the conducting elements (Bowman and Stroud 1985, 1989).

The enhancement factor of the field near a long defect (made of n C_0 capacitors) perpendicular to the electrodes is given by

$$E_m = E(1+kn), \qquad (2.64)$$

as above (see eqn (2.8)). Note that in the present case this expression is good in all dimensions.

The probability to find a long defect made of n C_0 capacitors is

$$P(n) \sim p^n L^2 \quad (2D), \qquad P(n) \sim p^n L^3 \quad (3D), \qquad (2.65)$$

and the value n_c of the size of the most probable defect is again

$$n_c = -\frac{d}{\ln p} \ln L, \qquad (2.66)$$

with d equal 2 or 3 following the dimension of the system. The field in the C_1 capacitor adjacent to the most probable defect is given by

$$E_m = E\left[1 + K_1\left(-\frac{\ln L}{\ln p}\right)\right], \qquad (2.67)$$

where K_1 is different in two or three dimensions. Since $V_b = E_b L$, we get that the breakdown voltage is given by

$$V_b = \frac{E_b L}{[1 + K_1(-\frac{\ln L}{\ln p})]}. \qquad (2.68)$$

The distribution function $F(E)$ giving the probability that the sample will break when a field E is applied is obtained by a procedure identical to that we used for the fuse model. The result is

$$F(E) = 1 - \exp\left[-AL^d \exp\left(-\frac{K}{E}\right)\right], \qquad (2.69)$$

where d is the dimension of the system.

(b) The limit p approaching p_c

In this limit, we consider clusters of conducting defects and their mean size is the percolation correlation length ξ. The voltage V_1 between two such neighbouring clusters is given by $V_1 = V\xi/L$. The field between two clusters is increased as the field is zero inside a conducting cluster. The maximum value will be between the clusters with minimal distance between them; i.e. one unit cell of the lattice. When the field between two such clusters reaches a threshold value e_c, the breakdown takes place. One can then write $V_1/a = e_c$, or $V_b/L = e_c a/\xi$, giving

$$E_b \sim (p_c - p)^\nu. \tag{2.70}$$

This result was obtained by Lobb et al. (1987).

This value of E_b can be seen only as a mean value, but not as the most probable value, since the cluster size can fluctuate around its mean value. What is necessary to identify here is the size of the most dangerous cluster. The voltage drop between two neighbouring clusters is proportional to their size l. The maximum voltage drop is across the most probable largest cluster. Using the same method as for the fuse problem, one finds $l_m \sim \xi \ln[(L/\xi)^d]$ for the size of the most dangerous defect cluster. Using a similar argument as above, one gets

$$E_b \sim \frac{(p_c - p)^\nu}{\ln(L/\xi)}. \tag{2.71}$$

This expresssion is valid if L is larger than the correlation length, since the infinite cluster is described as a super-lattice with a unit cell size equal to ξ. Since ξ appears in the logarithm, it is possible to neglect it. In such cases, the most probable value of the breakdown field is

$$E_b \sim \frac{(p_c - p)^\nu}{\ln L}, \tag{2.72}$$

as proposed by Beale and Duxbury (1988).

As above, we recall the remark by Bergman (Bergman and Stroud 1992), that very near p_c, E_b may become independent of L, as some cross-over (to percolation dominated statistics) is expected.

Also, this result is valid independent of the dimension of the system. As mentioned above, this is due to the shape of the breakdown path, which is always, on the average, perpendicular to the electrodes. However, we recall that the correlation exponent is different in two or three dimensions.

(c) Brief summary

As in the fuse model, one can summarise all the results in one formula analogous to the formula (2.29).

The formula for the breakdown electric field is

$$E_b = E_0 \frac{[(p_c - p)/p_c]^\phi}{1 + K \left[\frac{\ln(L/\xi)}{(-\ln p)}\right]}. \tag{2.73}$$

This formula is valid in all cases, with the exponent ϕ ($= \nu$ for lattice percolation) dependent on the dimension and on the type of percolation (see below the results for continuum percolation).

The dielectric breakdown problem

Table 2.2 Theoretical estimates for the dielectric breakdown exponent t_b

Dimension (d)	Lattice percolation	Continuum percolation
2	ν 1.33	$\nu + 1$ 2.33
3	ν 0.9	$\nu + 1$ 1.9

2.3.4 Continuum percolation

In the present case, one introduces conducting defects in the form of circles in two dimensions or spheres in three dimensions with randomly positioned centres having the possibility to overlap. The size of one defect is the characteristic length of the problem, as mentioned in the case of the fuse problem.

We shall only consider the limit p very near to p_c when one can use the results of percolation. In this case, if the minimum possible distance between the clusters is $\delta_{\min}$ then, following the same argument as above, one gets $E_b \sim \delta_{\min}/\xi$, or

$$E_b \sim (p_c - p)^{\nu+1}. \tag{2.74}$$

To conclude we recall that the formula (2.73) is still valid in the present case, and the values of the exponent ϕ ($= t_b$ here) are given in Table 2.2.

2.3.5 The shortest path

We mentioned this concept when we introduced the electromigration fuse model (Section 2.2.5). We recall it in the present context of the dielectric breakdown problem. Above we considered a walker which jumps from bond to bond since the problem was to break bonds until failure. Here, we adopt a slightly different definition specific to the present problem.

The walker begins its walk from one electrode with jumps from one site to another and reaches the opposite electrode after executing a self-avoiding walk. We suppose that the walker can transform an insulating bond into a conducting one, when jumping between two sites separated by an insulating bond. After completion of the walk, the dielectric sample is broken since it creates a continuous conducting path between the electrodes.

Let n_0 denote the number of insulating bonds transformed into conducting bonds by the walker during its walk across the sample. The shortest path for a given configuration is the path with the smallest n_0. And when averaged for a large number of configurations, one gets the mean value of n_0, $g = \langle n_0 \rangle / L$.

The interest of the concept of shortest path is evident in the case of the dielectric breakdown. In the electromigration fuse model, it was useful only in two dimensions, but here it can be of interest in all dimensions.

An extensive study of the shortest path for the dielectric problem was made by Stinchcombe et al. (1986), Duxbury, Shukla, Stinchcombe and Yeomans (1987) in an unpublished paper, and Duxbury and Leath (1987). These authors use a different terminology and call the shortest path the minimum gap. They determined $g(p)$ for all the range of p (below or equal to p_c), in two dimensions (square lattice) and in three dimensions (simple cubic lattice) for regular and directed percolation. In directed percolation, the current through a conducting bond is possible only in one direction, and in the lattice, all the vertical and all the horizontal conducting bonds are directed in the same sense (for example from left to right for the horizontal bonds and from the bottom to the top for the vertical bonds).

In all these four cases, they found that g goes to zero at p_c with an exponent equal to that of the correlation length. For p approaching 0, g goes to unity, but for the three-dimensional regular percolation and for the directed percolation (in both $d = 2$ and 3) there is a steep decrease of g from unity when p begins to increase. This behaviour is reminiscent of the behaviour of E_b but this is not very well understood for the gap.

2.3.6 Dielectric breakdown with tunnelling bonds

Recently, Sen and Kar-Gupta (1994) and Kar-Gupta and Sen (1995) proposed a new percolation model to mimic the properties of nonlinear composites. However, it can also be interpreted as a special model for the dielectric breakdown problem.

They considered a lattice with a fraction p of the bonds having linear (or ohmic) resistance. These are called the o-bonds. Among the insulating bonds, some are identified as the tunnelling or t-bonds: an insulating bond located between two nearest-neighbour o-bonds is a t-bond. A t-bond is insulating if the voltage across it is smaller than a pre-assigned threshold value, and becomes a linear resistor for voltages larger than the threshold. The other insulating bonds remain insulating, and can not be broken. One can now define two kinds of (geometric) percolation thresholds for such random networks. One is the normally defined percolation threshold p_c for the o-bonds. The other is the minimum concentration p_{ct} ($< p_c$) of the o-bonds at which the network of all the o-bonds and the corresponding t-bonds forms a system spanning cluster. For a random network of such bonds, under external electric field E, one can have the following three scenarios:

(i) If the concentration p of the o-bonds is below p_{ct}, then even if the applied field is large enough to ensure all the t-bonds become conducting,

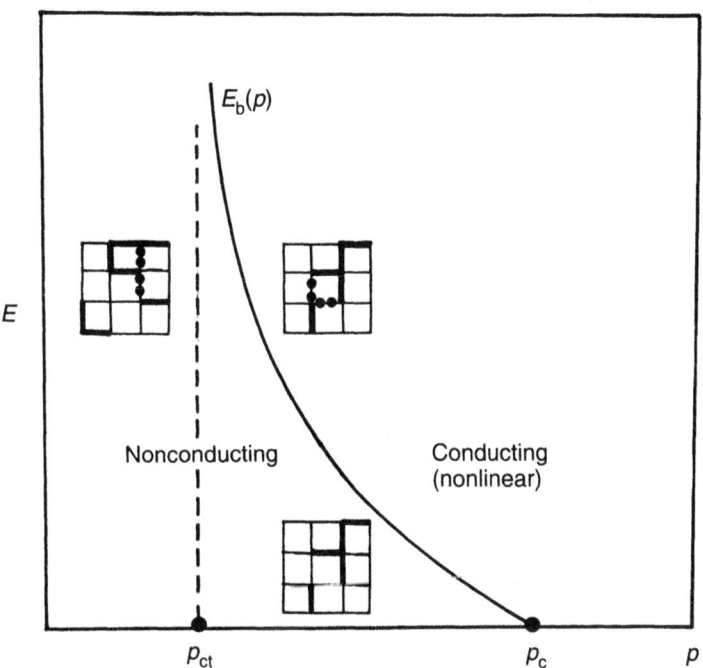

FIG. 2.14. Schematic variation of the dielectric breakdown field $E_b(p)$ with the o-bond concentration p, in the ohmic bond-tunnelling bond networks. The typical contributions of the t-bonds are indicated in the three insets; when the t-bonds become conducting, they are indicated by the dotted lines.

the system is still an insulator.

(ii) If p is larger than p_{ct} but smaller than p_c, the sample is insulating at low fields (voltages) as the t-bonds are still insulating. However, for larger fields, the t-bonds are transformed into conductors and the network of the t- and o-bonds forms a spanning cluster.

(iii) For $p > p_c$, the system is already percolating through the o-bonds, and is obviously a conductor. However, since the t-bonds between the dangling ends of the infinite cluster of the o-bonds can become conducting, adding to the net conductivity of the network, the resultant conductivity of the network becomes nonlinear. This case is useful for modelling the nonlinear conductivity of composite materials (Gefen et al. 1986, Bardhan and Chakrabarti 1994, Sen and Kar-Gupta 1994).

In the case (ii), when $p_{ct} < p < p_c$, there is clearly a breakdown voltage or field E_b, beyond which the sample becomes conducting. The difference with the regular dielectric breakdown problem (discussed earlier) is that here only the t-bonds can be broken, as they are located between two conducting bonds. This model can thus simulate the behaviour of nonlinear conductors with hopping or tunnelling between the ohmic parts of the sample. Kar-Gupta and Sen (1995) made a numerical estimate for the p_{ct}, which is about 0.18 for square lattice ($p_c = 0.5$). They also estimated the breakdown field E_b as a function of p numerically. E_b diverges for p near p_{ct}, and vanishes at $p = p_c$ (see Fig. 2.14, where some typical operations of the t-bonds in different regions are also indicated in the small insets). The breakdown field $E_b(p)$ is found to have the scaling behaviour

$$E_b(p) \sim (p_c - p)^{t_b}$$

near p_c. The numerical estimate of t_b ($\simeq 1.4$ for a square lattice) is not inconsistent with $t_b = \nu$, the normal percolation correlation length exponent.

2.3.7 Numerical simulations and experimental results

(a) Numerical simulations

Several groups have performed numerical simulations in the square lattice and the simple cubic lattice, and the results of these simulations are in complete agreement with the above theoretical predictions.

Bowman and Stroud (1989) determined E_b in both two and three dimensions, and for both bond and site percolation near the percolation threshold. Since we have not discussed the study of breakdown problems in site percolation so far, we shall give some words here on this situation. In site percolation, all the bonds are conducting but can or cannot be connected with other neighbouring bonds, depending on the state of the sites. A site can be insulating and the adjacent bonds are not connected, or a site can be conducting and the adjacent bonds are connected. If p, the amount of the conducting sites, is larger than p_c (where there is an infinite cluster of conducting sites) the sample is conducting.

Bowman and Stroud (1989) solved numerically the Laplace equation at all the sites of the lattice made of insulators (with probability $1 - p$) and conductors (probability p) and got the values of the potential across each bond. This is solved by taking into account the condition that no potential difference can be maintained within a conducting cluster. They found that E_b (defined as the field giving the breakdown of the first bond) decreases towards 0 as p approaches p_c, with an exponent equal to 1.1 ± 0.2 in $d = 2$ and equal to 0.7 ± 0.2 in $d = 3$ for both site and bond percolation. These values are consistent with the above predictions, namely that the exponent of E_b must be equal to the correlation length exponent ν. We recall that

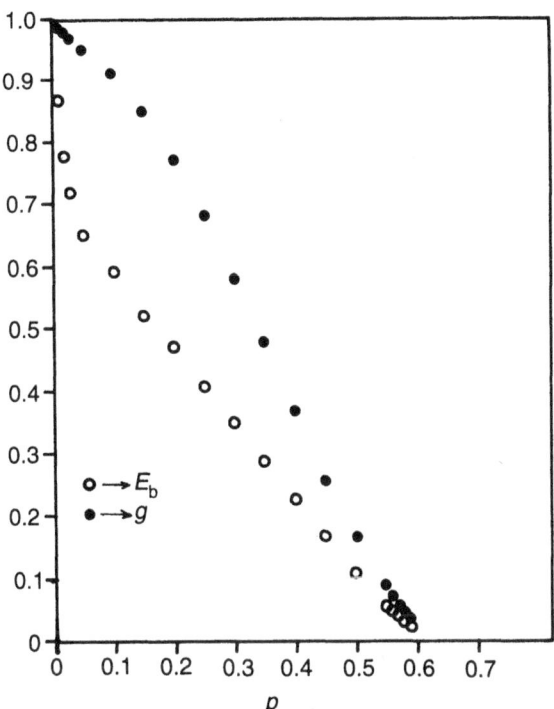

FIG. 2.15. Failure field (o) and shortest path or gap (•) on a square lattice. Computer simulation by Manna and Chakrabarti (1987). The two quantities behave differently near $p = 1$ but have a similar variation near p_c.

ν is equal to 4/3 and 0.9 in $d = 2$ and 3 respectively. The values found by Bowman and Stroud are slightly smaller than the theoretical values and this is a very common result in simulations and experiments since it is not possible to go very near p_c.

Manna and Chakrabarti (1987) performed the same kind of calculation for site percolation in a square lattice for the entire range of p $(0 < p < p_c)$ and at the same time determined the minimum gap g (or the shortest path). These results are shown in Fig. 2.15. Near p_c, they found that both E_b and g go to zero with almost the same exponent value equal to about unity. This is still consistent with the above theoretical analysis. We point out that Bowman and Stroud worked with $L = 150$, while the results of Manna and Chakrabarti are for $L = 25$ only (both in $d = 2$). The smaller size of the sample limits the possibility to reach values of p near enough to p_c and

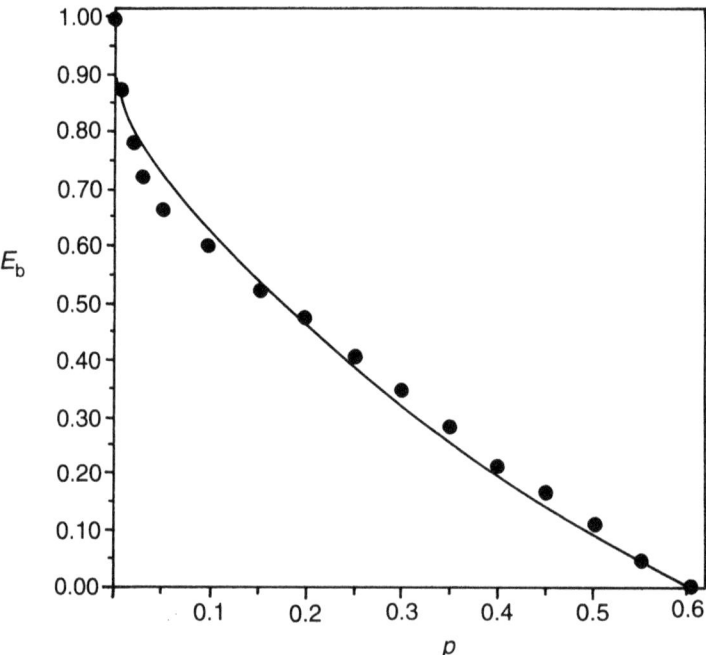

FIG. 2.16. Fit of the results of Manna and Chakrabarti (Fig. 2.15) with the expression (2.73); $\phi \simeq 1.0$.

this explains the results of Manna and Chakrabarti.

Beale and Duxbury (1988) checked the dependence of the breakdown field on the sample size. First they found a linear dependence of $1/E_b$ with $\ln L$. They fitted the results for $E_b(p, L)$ with the form

$$E_b = E_0 \frac{1}{A(p) + G(p)\ln L}. \qquad (2.75)$$

The exact expressions for $A(p)$ and $G(p)$ are not known in the entire range of p and (2.75) is not strictly equivalent to (2.73), which can be seen only as an interpolation formula. Qualitatively, it is found that $G(p)$ and $A(p)$ increase when p increases towards p_c. Near p_c, one has $A(p) \sim (p_c-p)^{-\nu}$ and $G(p) \sim (p_c - p)^{-\nu}/\ln p$. Beale and Duxbury (1988) determined $G(p)$ near p_c, and by plotting $\ln(G(p)\ln p)$ versus $-\ln|p-p_c|$, it is found that the value of the correlation length exponent ν is around 1.15, in good agreement with the theoretical value. Bowman and Stroud (1989) also studied the above dependence of E_b on $1/\ln L$ at p values very close to p_c, and found its validity even at $p = p_c$. Of course, the quality of their data here is not very good, because of large fluctuations.

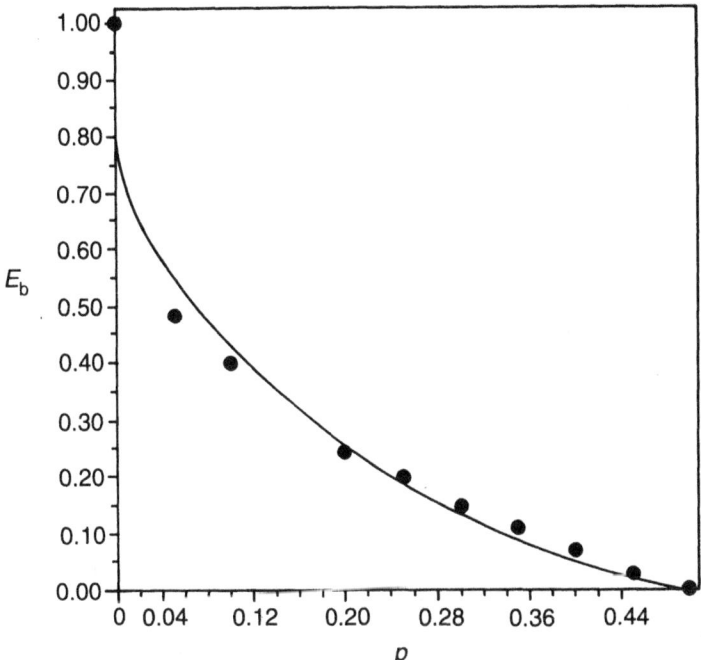

FIG. 2.17. Fit of the computer simulation results of Beale and Duxbury (1988) with the expression (2.73); $\phi \simeq 1.2$.

Beale and Duxbury also determined the cumulative failure distribution probability $F(E_b)$ and found very good agreement with the Gumbel double exponential form (2.69).

Considering now the variation of E_b with p, we compare the results of Manna and Chakrabarti (1987) and of Beale and Duxbury (1988) with the formula (2.73). We show in Fig. 2.16 and Fig. 2.17 how these numerical results compare with the expression (2.73). As one can see, there are satisfactory agreements. The magnitude of the exponent ϕ is found to be about 1.0 from the results of Manna and Chakrabarti, and around 1.2 from the results of Beale and Duxbury. Here also it is better to accept this exponent as an effective one.

(b) Breakdown susceptibility

Acharyya and Chakrabarti (1995, 1996a) presented recently a study of the breakdown process by defining a breakdown susceptibility. We recall the procedure used by de Arcangelis et al. (1985), Duxbury et al. (1987) and Beale and Duxbury (1988) in order to produce the failure path. First, one determines the voltage V_b necessary to break the first bond. This bond is removed (replaced by a conductor in the dielectric breakdown case) and the voltage necessary to break the second bond(s) is determined, and so on until one reaches the voltage necessary to break the last bond. For a particular initial configuration, although there are occasional increases, one generally gets a series of decreasing voltages for the breaking of successive bonds. In general, the complete breakdown field is observed here to be slightly larger than the initial breakdown field (Beale and Duxbury 1988).

The procedure of Acharyya and Chakrabarti is different. After applying V_b and replacing the broken bond by a conductor, a larger voltage is applied and the broken bonds are localised and then replaced by conductors. The process is continued until one reaches the voltage needed to break the last bond. When one reaches this final breakdown voltage V_b^{fin} ($= E_b^{\text{fin}} L$), a continuous (percolating) path of broken (or conducting) bonds exists. The process can be continued further until all the bonds in the sample are broken if V is made really large enough. After averaging over the initial random configurations (for fixed p), one can define a function $n(V)$, giving the average number of broken bonds (broken dielectric bonds which become conducting) in the sample for a given externally applied voltage V. Clearly, for very large values of V, n goes to L^2 or L^3 depending on the dimension. Since $n(V)$ is an increasing function of V and eventually saturates, it exhibits an inflection point. The question is to determine its position. Here, one can define the breakdown susceptibility $\chi = \mathrm{d}n/\mathrm{d}V$, which has a maximum at this special point. In fact, the statistics of these growing broken clusters here is quite intriguing, and has been studied by Acharyya *et al.* (1996).

Acharyya and Chakrabarti studied the behaviour of this breakdown susceptibility, using numerical simulations in two dimensions. They determined $\chi = \mathrm{d}n/\mathrm{d}V$ as function of V for numerical simulations of site dilute lattices, and found effectively that χ has a maximum at a voltage V_b^{eff} different from V_b^{fin} (see Fig. 2.18). However, with increasing sample size, they found that (a) the voltage V_b^{eff} approaches towards V_b^{fin}, and (b) the maximum value of χ at this V_b^{eff} itself increases with the system size L. So, the conclusion of this study is that for an infinite sample the breakdown susceptibility is very likely a divergent quantity and this divergence occurs for the voltage at which a path of broken and conducting bonds appears. This opens the way to a possibility to predict the value of V_b^{fin}, without going to the complete breakdown point of the sample.

(c) Distributed breakdown threshold

In the lattice model of random dielectrics, we have considered, so far, dielectric bonds with fixed breakdown threshold. We now consider the problem of breakdown of a random dielectrics, where the bonds have random break-

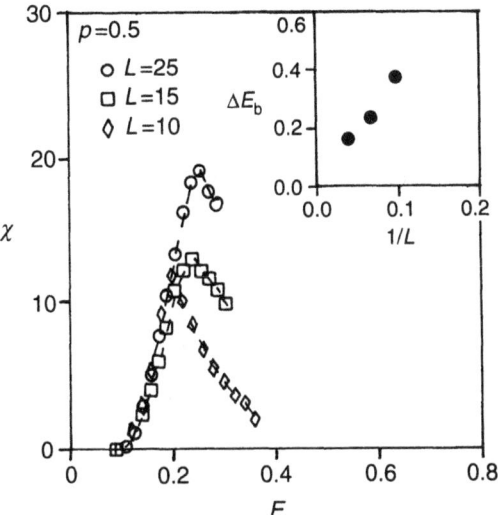

FIG. 2.18. Breakdown susceptibility as a function of the applied voltage in the dielectric breakdown problem for several sample sizes (from Acharyya and Chakrabarti 1996a). The inset shows that ΔE_b, the difference of E_b estimates from the peak position in χ and from the percolation of the broken cluster, decreases with increasing system size.

down thresholds. One generally considers here the transport properties of such a lattice of random dielectrics, which become resistors beyond their respective threshold voltages:

$$i = g(v - v_c) \quad \text{for} \quad v > v_c,$$

and

$$i = 0 \quad \text{for} \quad v < v_c, \qquad (2.76)$$

where i denotes the current through the bonds for a voltage drop across it beyond a value v_c, which is different from bond to bond and is chosen to be distributed uniformly between 0 and 1 (Roux et al. 1987, Roux and Herrmann 1987, Sahimi 1995).

If the applied voltage V on the sample is low, the sample will not be conducting. Beyond some breakdown voltage V_b^c, the sample begins to conduct non-linearly. Finally for large enough voltage, all the bonds are conducting and the sample becomes a linear conductor. One can relate this problem to the dielectric breakdown problem considering V_b^c as equivalent to the breakdown voltage V_b^{fin} ($= E_b^{\text{fin}} L$) defined above, since a percolating conducting path exists in each cases.

From numerical simulations, it is found that V_b^c is about 0.30 in a square lattice and about 0.17 in a simple cubic lattice (Sahimi 1995, Rosen and Mamun 1993).

(d) Experimental results

We shall describe two experimental determinations of the breakdown field: one in three, and the other in two dimensions.

Coppard et al. (1989) measured the breakdown field of polyethylene plaques loaded with metallic particles in very small quantities. Their results concern the range of low p. One of the purposes of this study was to determine the distribution probability $F(E_b)$: to distinguish between the Gumbel and the Weibull distributions. But it appears to be very difficult a task experimentally, because a large number of samples are needed for the study. However, they were able to verify that the average of the breakdown fields varies with p in accordance with (2.68).

Benguigui (1988) and Benguigui and Ron (1994) performed experiments in two dimensions on a square lattice of random resistors and light-emitting diodes (LED). The LEDs have the following property: there is a dramatic change in their resistance beyond a well-defined voltage v_b of about 1 volt. Below this voltage, the diode has a very large resistance (about $10^7 \Omega$) and above this voltage a small one ($10^3 - 10^2 \Omega$). It is possible to see such a diode as a device which is a quasi-insulator for $v < v_b$ and a conductor for $v > v_b$. Such a device can mimic the breakdown of an insulator as in the dielectric model. The transition between the two states is relatively sharp, such that once a diode is in the conducting state (representing a broken insulator) the voltage remains constant. This is the main difference with the model dielectric bonds discussed earlier, where the voltage across the dielectric bonds becomes zero after breakdown. However, the advantages of using LEDs to model breakdown of dielectrics are obvious: their behaviour is reversible, i.e. the LED comes back to its high resistance state once v is decreased below v_b, and the breakdown process becomes visual as the light is emitted in the conducting state of the diode.

Consider now a lattice with resistors (with probability $p < p_c$) and LEDs (with probability $1 - p$). A voltage is applied across such a square lattice. For low voltages, all the LEDs are in their high resistance state and since $p < p_c$, the sample is practically nonconducting and has a large resistance. Increasing V, one finally gets a dramatic increase in the current flowing through the sample (by several orders of magnitude) at a well-defined value V_b of V. This voltage is taken as the breakdown voltage of the sample. In Fig. 2.19, the variation of V_b with p is shown, and one sees that V_b goes to zero for $p = p_c$. A detailed study of the variation of V_b near p_c shows that V_b is proportional to $|p - p_c|^{1.1}$.

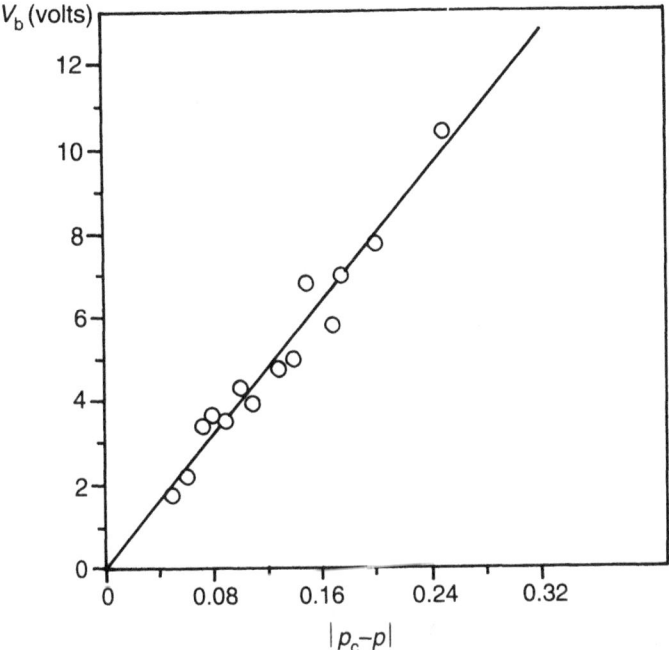

FIG. 2.19. Failure voltage V_b versus $(p_c - p)$ in the laboratory simulations of breakdown in random dielectrics with LED. The full line is $(p_c - p)^{1.1}$.

When a conducting path develops at $V = V_b$, this voltage is proportional to the number of LEDs belonging to this connecting path. Since it is the smallest voltage giving a conducting path, this path is the shortest path defined in Section 2.3.5. In Fig. 2.20, the continuous line indicates the shortest path as was determined from the numerical simulations (Duxbury et al., unpublished). The agreement is indeed very good. The exponent value 1.1 ± 0.05, for V_b variations near p_c, is very near that (ν) for the correlation length exponent or for the shortest path. As in other measurements and simulations, the experimental value is smaller than the theoretical value because of the small sample size ($L = 20$ in the present case).

Another interesting result of this study is the determination of n, the number of 'broken' LEDs in the breakdown path as a function of p. There are several examples of such study using computer simulations, where only near p_c, an effective exponent with value slightly smaller than ν was found. Benguigui and Ron determined $n(p)$. They found that near p_c, n decreases with an exponent value near 1.1. However, they also observed that n ex-

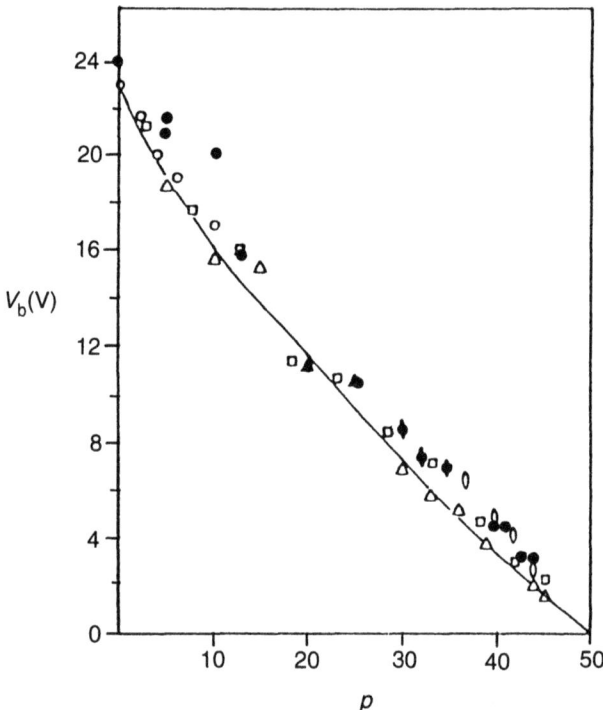

FIG. 2.20. Failure voltage V_b for the whole range of p in the dielectric breakdown simulation experiment with LED. The full line denotes the shortest path (normalised with the V_b value at the point $p = 0$) as determined by Duxbury et al. (unpublished). Different symbols indicate different realisations of the sample (Benguigui and Ron 1994).

hibits a maximum at about $p = 0.22$-0.25. We do not have any explanation for this observation. For more details on the shape of the conducting path, we refer the reader to the original papers.

2.4 Conclusions

From the discussions in the preceding sections, we think it is clear that the problem of the first failure (fuse or breakdown) is understood quite well. This first failure is described by the quantity I_f or V_b. We believe that the parameters controlling these quantities are now well-known. All the numerical simulations and the experiments (even though they are rather scarce), give a complete confirmation of the theoretical developments. In particular, the experiments of Yagil et al. (1992, 1993) give important confirmations for the behaviour of the fuse current I_f of random fuse networks, although the behaviours observed for large currents do not necessarily conform with what is expected.

The important open question is precisely what happens just after the first bond is fused or broken. There is no precise answer, and it seems that the efforts in the future will be concentrated in this direction. We reported that the current belief is that in the lattice models, after the failure of the first bond, a cascading effect occurs and the failure propagates, even when the current or the voltage across the sample is kept constant at the first failure value. However, this dynamic problem has not been solved yet. Very recently of course Zapperi et al. (1997) have confirmed the critical divergence of breadown susceptibility χ at the breakdown point in the random fuse some other models, as discussed in the section 2.3.7(b). They also argued that this divergence of $\chi \equiv \int nP(n)dn \sim (V_b^{fin}-V)^{-\gamma}$ suggests a power law $P(n) \sim n^{-\tau} f(n[V_b^{fin} - V]^k)$, with the scaling function f defined asymptotically and $\gamma = (2-\tau)k$, for the avalanche size distribution $P(n)$. This is very similar to the Guttenberg–Richter law for the earthquake magnitude distribution (see section 1.2.3 and also chapter 4).

Even when the above dynamic problem is replaced by a series of (quasi-) static problems, one does not have much information. One question is the exact shape of the failure path. It is probably a fractal, but the fractal dimension has not been investigated. Two procedures have been been reported for the formation of the failure path: one with decreasing voltages and another with increasing voltages, and it is not known if these two procedures will give the same failure path. A third procedure, when the the external current or voltage remains constant at I_f or V_b and one removes the broken bond, can be considered. It should be very useful to compare these three procedures, and the differences in the propagation or growth characteristics.

Another direction is to build models which will be more realistic. In all these models described above, the exact microscopic mechanism of the failure was not considered. But it seems very likely that the exact nature of the process will influence what happens after the first failure. Yagil et al. (1992, 1993) observed that after the first failure (fuse), the resistance of the sample can get decreased or increased depending on the failure process. If increase is what one expects, then the decrease means that the first failure improves the contact between the parts which melt. Thus, only by a detailed analysis of the failure process can one understand it. To come back to the dynamic problem, it is also very likely that the velocity of the failure propagation will depend on the failure mechanism.

3

FRACTURE STRENGTH OF DISORDERED SOLIDS

3.1 Introduction

For many engineering applications, an important aspect of the choice of the load-bearing material and the engineering design is that the solid material should not fracture or fail in service. As the material contains natural disorders or defects, produced during the preparation or synthesis of the solid, solids contain built-in microcracks (where the interatomic interactions become vanishingly small due to large separation distances etc.) produced due to the defects. Fracture properties of the solids containing these microcracks, under different loading conditions, are therefore of extreme importance in civil and other engineering design problems. We will discuss the statistical aspects of the fracture properties of such solids in this chapter. There are three modes of loading a solid, as shown in Fig. 3.1, where the arrows denote the stress directions. We will mostly discuss here the fracture strength of randomly disordered solids under stress or loading condition in mode I. However, because of the mixing of tensile and shear stresses in disordered systems, the results discussed in this chapter are, in effect, valid for all the modes.

As in the case of electrical failure in random conductor-insulator networks in the earlier chapter, we first discuss here the concept of stress concentration in an otherwise perfect solid, which is stressed and contains a single crack inside. Here, the stresses concentrate at the sharp edges of the crack, where it can become much larger compared to the external force. As one increases the external force, the crack starts propagating from such

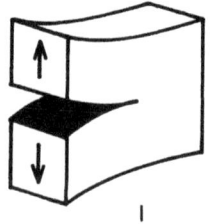
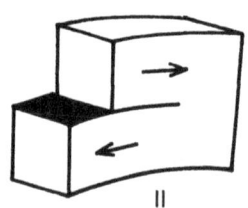

FIG. 3.1. Three modes of loading a solid sample.

sharp edges, where the 'concentrated' stress goes beyond the threshold limit for the material. Since no deformation (in the shape of the crack at its tips) occurs for brittle solids, the solid becomes even more weaker with the extended crack as it propagates, and the entire solid fails. For ductile solids, extensive deformation due to stress concentration around the sharp edges of the crack often reduces the sharpness of its edges and hence the stress concentration. This often arrests the fracture propagation, and some (transverse) growth of the initial crack inside the solid does not necessarily make the solid weaker. Thus the final failure stress of such a solid is usually higher than the stress at which the first breaking (or nucleation of fracture) occurs. It may be mentioned here that, for reasons of simplicity in theoretical modelling, we will mostly discuss here the fracture properties of brittle solids.

We discussed above the case of a solid with a single crack inside. We will mostly concentrate in this chapter on the statistical properties of fracture in randomly disordered brittle solids. As in the electrical breakdown case, we model the random solids using the percolation models (lattice or continuum), where the initial random cracks are modelled, at a semi-macroscopic level, by the vacancy clusters formed due to fluctuations. Stated more explicitly, we consider a lattice of which each bond represents a spring (which generally resists changes of its length as well as the inter-bond angle between the neighbouring bonds), and a fraction $1 - p$ of such bonds are randomly absent (or the corresponding springs are cut or broken).

The problem of modelling a solid by a discretised network of springs is actually quite nontrivial, contrary to the cases of electrical systems discussed in the previous chapter. If one considers a network made of regular springs (with the elastic energy contributed by the changes in the lengths of the springs), the network may be stable or unstable against shear deformations, depending on the lattice. For example, in two dimensions, the square lattice is unstable but not the triangular lattice. Thus, to increase the stability of the lattice, and to make its elastic properties more similar to those of continuous solids, bond-bending forces are added. Due to these bond-bending forces, the elastic energy changes when the angle between the neighbouring springs (connected at a site) changes, even if the lengths of the springs remain unchanged. However, even then, all these stable lattices can not completely mimic a solid, in that their Poisson ratios can not take all the values in the regular range $(0, 1/2)$. For example, in the triangular spring network, the Poisson ratio is $1/3$. An interesting discussion of this problem can be found in Jagota and Benisson (1994).

We study the breaking properties or the fracture strength, and its statistics, for such model solids. The initial cracks, present before the solid is stressed, are now more than one and are no longer linear. They have var-

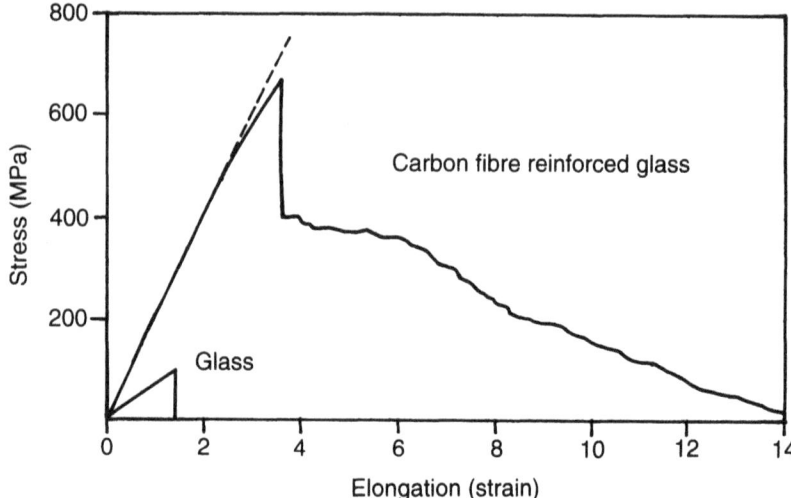

FIG. 3.2. Typical breaking characteristic for (brittle) glass and (ductile) carbon fibre reinforced glass (from Herrmann and Roux 1990).

ious shapes and sizes, of which the statistics is given by the percolation cluster statistics discussed earlier in Section 1.2.1 of Chapter 1. When such a solid is under stress, one again needs to identify the most dangerous or weakest (often the biggest) crack, which fails first due to maximum stress concentration at its edges. As discussed earlier (in Section 1.2.2, and also in the previous chapter), the extreme nature of the statistics for such dangerous clusters or cracks involve the tail portion of the percolation cluster distribution function (for the large clusters) and give rise to the Weibull or Gumbel-like fracture stress distribution functions $F(\sigma)$.

We will discuss these fracture properties of disordered solids, modelled by the random percolation models, and concentrate on their statistics, given by the cumulative failure strength distribution $F(\sigma)$ under stress σ, and the most probable fracture strength σ_f of such samples. We will discuss separately the cases for weak disorder ($p \simeq 1$) and strong disorder ($p \simeq p_c$). The scaling properties of σ_f near $p \simeq p_c$, and the nature of the competition between the percolation and extreme statistics here, will be discussed in detail.

3.2 Fracture strength of a perfect solid containing a single crack: Griffith's law

In an ideal solid, there are various methods of calculating the moduli of elasticity Y, from the knowledge of interatomic potentials and the lattice

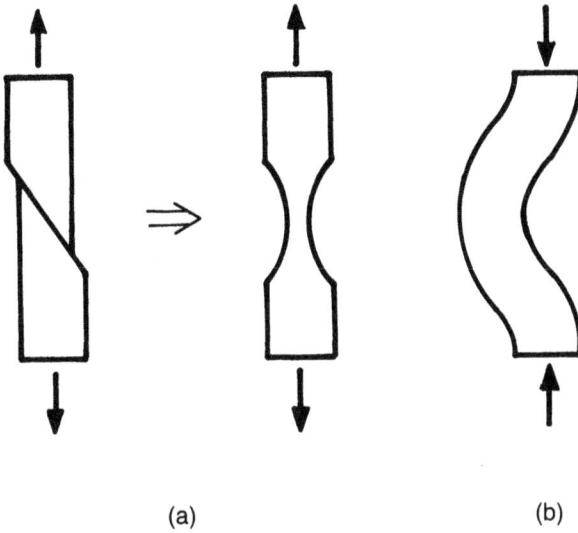

FIG. 3.3. Structural instabilities in ductile materials: (a) necking and (b) buckling.

structure (Born and Huang 1956). If one tries to calculate the fracture strength from the knowledge of these interatomic potentials, one gets a theoretical value which is typically orders of magnitude higher than the observed value in real specimens (see e.g. Marder and Fineberg 1996).

It has long been known that the fracture strength of a specimen is not any intrinsic material property; the fracture strength of a given material fluctuates considerably (even by an order of magnitude) and is not generally very reproducible. Also, the brittle solids like glass, which remain elastic (linear) up to the fracture breaking point, are found to fail suddenly while the ductile materials, like the metallic solids, deform extensively by plastic flow prior to rupture or failure of the specimen (see Fig. 3.2). The linear (and reversible) elasticity without any significant plastic deformation up to the point of fracture of brittle solids is indicated, for example, by the fact that the broken glass or ceramic pieces can be rejoined edge by edge using synthetic adhesives. Unlike such brittle materials, where very little deformation take place before the fracture, the ductile materials deform considerably (by necking or buckling due to dislocation slips in crystal planes (see e.g. Fig. 3.3a and b) before fracture. It has, in fact long been identified (from these observations) that the thin microcracks within solids, coming from random vacancies, dislocations etc., are responsible for such easy fracture or breakdown of real solids.

By the first decade of this century it was established that material failures occur at such low stress levels, because real materials do not usually have a perfect crystalline structure and almost always some vacancies, interstitials, dislocations and different sizes of thin microcracks (having linear structure and sharp edges) are present within the sample. Since the local stress near a sharp notch may rise to a level several orders of magnitude higher than that of the applied stress, the thin cracks in solids reduce the theoretical strength of materials by similar orders, and cause the material to break at low stress levels. The failure of such (brittle or ductile) materials was first identified by Inglis (1913) to be the stress concentrations occurring near the tips of the microcracks present within the sample.

It may be mentioned here that although the defects and the consequent existing (micro-) cracks in the stressed solids are often microscopic in nature, except for cases with very high degree of disorder as in percolating solids (to be discussed later), the theoretical analyses of Inglis (1913) and Griffith (1920) are at the macroscopic level, using elasticity theory. Because of the linearity of the response strain to the applied stress in brittle solids, up to the rupture or breakdown point, quantitative theories for fracture were developed first for such solids.

3.2.1 Stress concentration

The analysis by Inglis showed that the local stress at sharp notches or corners of the microcrack can rise to a level several times that of the applied stress. This shows how the microscopic cracks or flaws within the solid might become potential sources of weakness of the solid.

For quantitative analysis, Inglis considered a uniformly stressed two-dimensional solid like a thin plate, containing an elliptic hole representing the crack (see Fig. 3.4). Let the lengths of the semi-major and -minor axes of the ellipse to be $2l$ and $2b$ repectively, and σ denote the external (say tensile) stress applied on the sample along the y-direction. We assume the (linear) Hooke law to hold everywhere in the plate and that the boundary surface of the elliptic hole, represented by the equation

$$\frac{x^2}{l^2} + \frac{y^2}{b^2} = 1,$$

is stress-free when l and b are small compared to the plate size.

We now intend to examine the modifying effect of the elliptic hole on the distribution of stress in the plate. A straightforward but tedious solution of the Laplace equation for the displacement vector field leads (See section 1.2.2(a), for the equivalent solution of the scalar potential problem in a two-dimensional conductor with an elliptic dielectric hole inside), to the largest concentration of stress at the point C, where

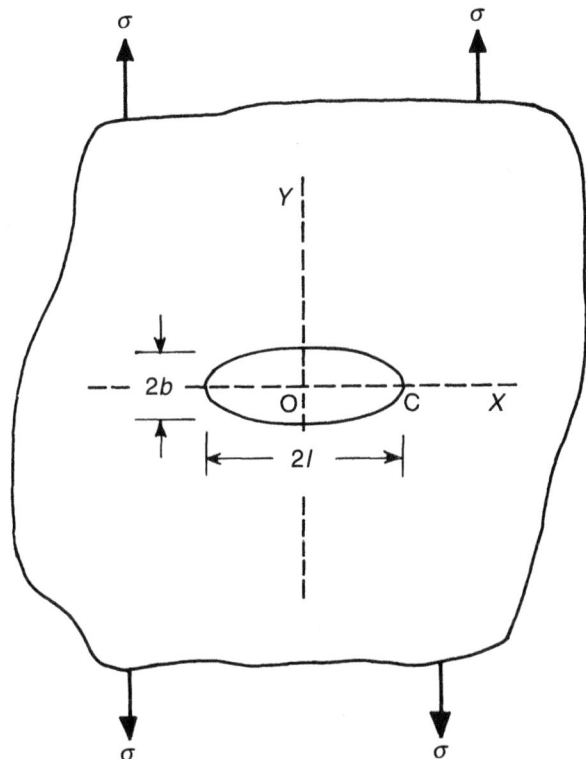

FIG. 3.4. A portion of a plate containing an elliptic hole with semi-axes lengths l and b, under uniform stress σ. The stress distribution within the plate becomes nonuniform near the defect (hole) and stresses concentrate at the tips (sharp edges at the horizontal ends) of the hole.

$$\sigma_{yy} = \sigma \left(1 + \frac{l}{b}\right)$$

$$= \sigma \left[1 + \left(\frac{l}{\rho}\right)^{1/2}\right], \tag{3.1}$$

where $\rho = b^2/l$ is the radius of curvature at the point C. A thin crack of length l can be represented by the ellipse in the limit $b \ll l$, when the stress concentration factor is given by

$$\frac{\sigma_{yy}}{\sigma} = \left(\frac{l}{\rho}\right)^{1/2}. \tag{3.2}$$

Usually the radius of curvature ρ at the sharp notch of the crack is determined by the atomic sizes and is very small. It is immediately evident that the stress concentration at the sharp notches of the microcracks can become extremely large due to the above stress intensity factor, and the fracture should start propagating from there. Although this analysis indicates clearly where the instabilities should occur, it is not sufficient to tell us when the instability does occur and the fracture propagation starts. This requires a detailed energy balance consideration.

3.2.2 Griffith's energy balance concept

Griffith in 1920, equating the released elastic energy (in an elastic continuum) with the energy of the surface newly created (as the crack grows), arrived at a quantitative criterion for the equilibrium extension of the microcrack already present within the stressed material. As before, we give below an analysis which is valid effectively for two-dimensional stressed solids with a single pre-existing crack, as for example the case of a large plate with a small thickness. Extension to three-dimensional solids is straightforward, and discussed later.

Let us assume a thin linear crack of length $2l$ in an infinite elastic continuum subjected to uniform tensile stress σ perpendicular to the length of the crack (see Fig. 3.5). Stress parallel to the crack does not affect the stability of the crack and has not, therefore, been considered. Because of the crack (which can not support any stress field, at least on its surfaces), the strain energy density of the stress field ($\sigma^2/2Y$) is perturbed in a region around the crack, having dimension of the length of the crack. We assume here this perturbed or stress-released region to have a circular cross-section with the crack length as the diameter (see e.g. Lawn and Wilshaw 1975). The exact geometry of this perturbed region is not important here, and it determines only an (unimportant) numerical factor in the Griffith formula. Assuming therefore half of the stress energy of the annular or cylindrical volume, having the internal radius l and outer radius $l + dl$ and length w (perpendicular to the plane of the stress; here the width w of the plate is very small compared to the other dimensions), to be released as the crack propagates by a length dl, one requires this released strain energy to be sufficient for providing the surface energy of the four new surfaces produced. This suggests

$$\frac{1}{2}\left(\frac{\sigma^2}{2Y}\right)(2\pi w l\, dl) \geq \Gamma(4 w\, dl).$$

Here Y represents the Young's modulus of the solid and Γ represents the surface energy density of the solid, measured by the extra energy required to create unit surface area within the bulk of the solid.

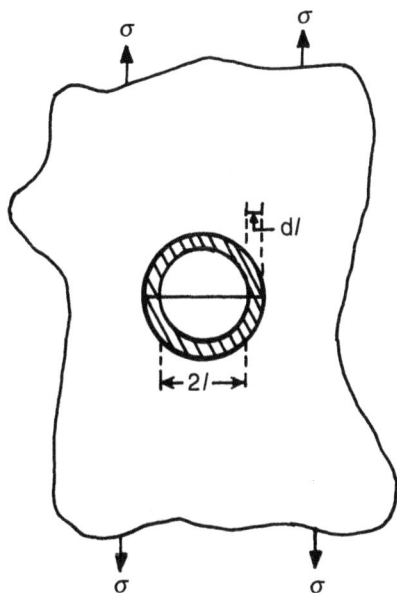

FIG. 3.5. A portion of a plate under stress σ (in mode I loading) containing a linear crack of length $2l$. For a further growth of the crack length by $2dl$, a fraction of the elastic energy of the shaded region may be assumed to get released.

We have assumed here, on average, half of the strain energy of the cylindrical region having a circular cross-section with diameter $2l$ to be released. If this fraction is different or the cross-section is different, it will change only some of the numerical factors, in which we are not very much interested here. Also, we assume here linear elasticity up to the breaking point, as in the case of brittle materials. The equality holds when energy dissipation, as in the case of plastic deformation or for the propagation dynamics of the crack, does not occur. One then gets

$$\sigma_f = \frac{\Lambda}{\sqrt{2l}}; \quad \Lambda = \left(\frac{4}{\sqrt{\pi}}\right)\sqrt{Y\Gamma}, \tag{3.3}$$

for the critical stress at and above which the crack of length $2l$ starts propagating and a macroscopic fracture occurs. Here Λ is called the stress-intensity factor or the fracture toughness. In fact, one can alternatively view the fracture occurring when the stress at the crack-tip (given by the stress intensity factor in (3.3)) exceeds the elastic stress limit for the medium.

In a three-dimensional solid containing a single elliptic disk-shaped planar crack perpendicular to the applied tensile stress direction, a straightforward extension of the above analysis suggests that the maximum stress concentration would occur at the two tips (at the two ends of the major axis) of the ellipse. The Griffith stress for the brittle fracture of the solid would therefore be determined by the same formula (3.3), with the crack length $2l$ replaced by the length of the major axis of the elliptic planar crack.

Generally, for any dimension therefore, if a crack of length l already exists in an infinite elastic continuum, subject to uniform tensile stress σ perpendicular to the length of the crack, then for the onset of brittle fracture, Griffith equates (the differentials of) the elastic energy E_l with the surface energy E_s:

$$E_l \cong \left(\frac{\sigma^2}{2Y}\right) l^d = E_s \cong \Gamma l^{d-1}, \qquad (3.4)$$

where Y represents the elastic modulus appropriate for the strain, Γ the surface energy density and d the dimension. Equality holds when no energy dissipation occurs and one gets

$$\sigma_f \cong \frac{\Lambda}{\sqrt{l}}; \quad \Lambda \cong \sqrt{Y\Gamma}, \qquad (3.5)$$

for the breakdown stress at (and above) which the existing crack of length l starts propagating and a macroscopic fracture occurs. It may also be noted that the above formula is valid in all dimensions ($d \geq 2$).

3.2.3 Experimental and computer simulational verifications of Griffith's law

Because of the difficulty in creating cracks of specific geometries within a solid, and also measuring their geometric properties, controlled microscopic experiments have not been possible to test the above formula (3.5) very accurately (Lawn and Wilshaw 1975, Thomson 1986). However, Griffith himself made several attempts to check his formula using some ingenious methods (Lawn and Wilshaw 1975), creating surface cracks on the solids and measuring their strength. Accurate studies of fracture have now been possible using computer simulations in modern computers.

(a) Griffith's experimental verification

Griffith attempted to check the validity of his formula for cracks on the surface of a solid. He took thin round tubes and spherical bulbs made of glass, and introduced (surface) cracks of fixed lengths (in the range 4 to 23 mm.) with a glass cutter. The specimens (with the above-mentioned cracks) were annealed prior to testing. The hollow tubes and bulbs were then burst by pumping in a fluid, and the critical stresses were then determined from the internal fluid pressure. The results were found to satisfy the relation $\sigma_f \sqrt{l}$ = constant, with a maximum scatter of about five per cent. This verified the essential validity of the formula (3.5). In fact, the application of the end loads to the tubes containing longitudinal cracks was found to have no detectable effect on the critical (fracture) conditions. This also confirmed that only the stress component normal to the crack plane is important.

(b) Computer simulation studies

With the advent of modern computers, the above theoretical investigations of pertinent atomic processes have been possible, using molecular dynamic simulations. The computer simulation results for lattice systems, with Lennard-Jones and other interatomic forces, support the above relation very well (Dienes and Paskin 1983, Paskin et al. 1980, 1981). Such numerical studies, although they might not be directly applicable to the actual engineering situations one faces in reality, helps otherwise in gaining considerable understanding of the fundamental processes involved in such phenomena. Also, unlike Griffith's theory, which is based on macroscopic elasticity theory, the computer studies can simulate directly the microscopic atomistic processes during the fracture propagation. As such, the computer simulation studies of fracture promise a very pure, straightforward and useful tool, though handicapped presently by the algorithmic bottlenecks of large scale molecular dynamic simulations.

In such molecular dynamic simulations, one starts with an array of 'atoms' or 'molecules', initially on a lattice, interacting with one another via an interatomic potential. These interacting potentials were taken by Paskin et al. (1980, 1981) to be the Lennard-Jones potential $\phi(r_{ij}) = \epsilon[(1/r_{ij})^{12} - 2(1/r_{ij})^6]$, where ϵ denotes the depth of the potential energy and r_{ij} denotes the interatomic separation of the atoms. This potential is assumed to have a sharp cut-off at an arbitrarily chosen value 1.6 (lattice constant) of the interatomic separation. The external stress or force is applied only at the boundary surface atoms. In order to investigate the Griffith fracture phenomena, one can consider for example a two-dimensional lattice of linear size L, remove a few ($l << L$) consecutive bonds along a horizontal row in the middle of the network, and apply tensile force on the upper and lower surface atoms in the vertical direction.

The molecular dynamics simulation essentially consists of solving Newton's equations of motion $d^2\mathbf{r}_i/dt^2 = \mathbf{F}_i$ for a suitably chosen small time interval for each of the atoms on a finite lattice. Here $\mathbf{F}_i$ denotes the resultant force contributed by the external source, if any, and all the interacting neighbours, through the respective $\phi(r_{ij})$'s on the ith atom. The mass of the ith atom or molecule is taken to be unity. The solution of the equation of motion may be obtained, using the Verlet scheme (see Allen and Tildesley 1987), writing the infinitesimal difference equation form of Newton's equation of motion as

$$\mathbf{r}_i(t + \delta t) = -\mathbf{r}_i(t - \delta t) + 2\mathbf{r}_i(t) + \mathbf{F}_i(\delta t)^2,$$

where δt refers to a suitably chosen small time interval. One solves these discretised equations for all the (d-) components of the displacement vectors $\mathbf{r}_i$, simultaneously for all the (L^d) nodes of the lattice and the equations become coupled as the position coordinates $\mathbf{r}_j$ for neighbouring lattice nodes appear in the force $\mathbf{F}_i$ at the ith node. Although the above dynamical calculations yield more and more accurate results as δt becomes smaller and smaller, it consumes more and more computer time to equilibrate. Choices of larger values of δt give rise to undesired oscillations in the positions of the atoms, and often lead to spurious results. These oscillations appear due to the fact that sequential solution of the equation of motion for any site i keeps the positions for the neighbouring atoms unchanged for the time period δt. As the restoring forces coming from the consequent displacements of the neighbouring atoms are not adjusted during the time interval δt, the ith site moves a larger distance than allowed. This is corrected for in the next iteration — when it moves back towards its equilibrium position but misses again for the same reason and shoots back further than required. This oscillation amplitude decreases as $\delta t \to 0$, which of course makes the program computationally time-intensive. In order to expedite the process of equilibration of the network, therefore, one often chooses a somewhat larger value for δt and introduces a small damping term which extracts some energy at each iteration and restricts the average speed of the atoms. For the calculation of $\mathbf{F}_i$ on each atom, one adds up those coming from the $\phi(r_{ij})$'s contributed by the occupied neighbours, as transmitted through the occupied or intact neighbouring bonds. Such neighbouring bonds may be initially cut or removed (to represent the initial microcrack), or are subsequently removed or broken during the process of dynamics (when the interatomic separation exceeds the threshold value chosen; the usual choice being 1.6 times the original lattice constant as mentioned above). The external force (σ_f) per atom, for which one additional (new) bond breaking occurs due to the increase in the interatomic separation beyond its threshold value, is taken is the fracture (nucleating) stress. This fracture nucleation stress is also the final fracture stress for such solids with one microcrack. This is because, at this stress level, the solid becomes even weaker with this increase in the crack length, and the fracture propagates across the sample.

Introducing a single linear crack of length l within triangular lattice networks of sizes $L \times L$ ($L \gg l$), Dienes and Paskin (1983) and Paskin et al. (1980, 1981) checked the validity of the Griffith formula and showed that $\sigma_f \sqrt{l}$ practically remains constant for small cracks. This observation of the validity of the Griffith formula, obtained using macroscopic (continuum elasticity) considerations, in such microscopic or atomistic simulation studies in computers, is quite intriguing. This, in fact, indicates the robustness of Griffith's criterion for the fracture propagation.

The above results are for regular spring networks, with a crack inside. As already mentioned in the introduction, such regular spring networks on square or simple cubic lattices (with such central force springs) are unstable against shear deformations. However, random spring networks (with random spring lengths) are stable against any deformation. The fracture properties of random spring networks have been studied numerically recently (Jagota and Bennison 1994).

3.3 Fracture strength of brittle solids with small disorder and rough (fractal) cracks

Except for perfect crystals, the fracture surfaces in solids are never very smooth. In fact, it seems now to be established experimentally that for weakly disordered solids, the surfaces formed during fracture processes are very rough, and on a mesoscopic scale (which is much larger than the atomic scale but smaller than the macroscopic sample size) they are observed to have self-affine properties (Mandelbrot et al. 1984, Bouchaud et al. 1993 a,b, Roux 1994). These have recently been observed using various fractographic investigations. By self-affinity of the fracture surface, we mean that the surface coordinate z in the direction perpendicular to the crack or fracture $(x - y)$ plane has the scaling property such that

$$z(\lambda x, \lambda y) \sim \lambda^\zeta z(x, y), \tag{3.6}$$

where ζ denotes the roughness exponent. Most of the recent experiments and their analysis suggest (Bouchaud et al 1993a,b) a universal value $\zeta \cong 0.80 - 0.85$ in three dimensions and $\zeta \simeq 0.67$ in two dimensions for the roughness exponent. In Fig. 3.6, we show a computer-generated rough self-affine surface with roughness exponent $\zeta \simeq 0.85$.

Such roughness of the crack surfaces induces some changes in the Griffith law for the strength of such weakly disordered solids. As the Griffith law is obtained by equating the released elastic energy with the surface energy, and the fractured surface here being rough, the Griffith law (3.5) gets modified for such cases. For a crack surface with roughness exponent ζ, the strength σ_f of a solid varies with the crack length l as

$$\sigma_f \sim \Lambda' l^{-\alpha}, \tag{3.7}$$

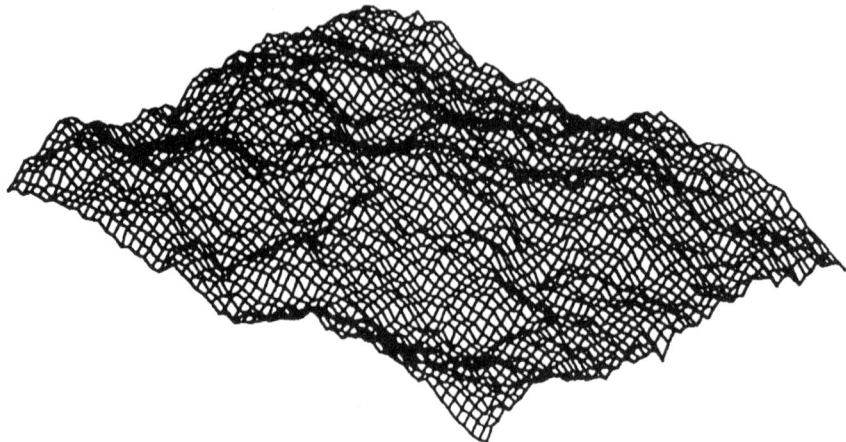

FIG. 3.6. A typical computer-generated crack surface, with the roughness exponent $\zeta = 0.85$ (from Roux 1994).

where $\Lambda' = \Lambda\sqrt{1-\alpha} = \Lambda\sqrt{\zeta/2}$, with $\alpha = (2-\zeta)/2$. This law of course reduces to the original Griffith law (3.5) for smooth crack surfaces with roughness exponent $\zeta = 1$. Or, more intuitively speaking, rougher cracks ($\zeta < 1$) are associated with larger and more singular stress fields.

3.3.1 Griffith's law for fractal crack surfaces

As mentioned above, the presence and occurrence of the rough crack surfaces (see Fig. 3.7) necessitates a change in the above Griffith estimate of the fracture strength or toughness (given by eqn (3.5)), where we had assumed a smooth crack surface. Let us express the stress concentration at the crack-tip generally by (3.7),

$$\sigma(l) \sim \Lambda' l^{-\alpha},$$

and represent the crack surface geometry by a self-affine fractal with roughness exponent ζ, up to a mesoscale of the order of ξ:

$$z \sim z_{\max}\left(\frac{r}{\xi}\right)^{\zeta}, \quad \text{for } r \ll \xi, \tag{3.8a}$$

$$z \sim \frac{z_{\max}}{\xi}, \quad \text{for } r \gg \xi, \tag{3.8b}$$

where ξ denotes the self-affine correlation length and $z_{\max}$ is the typical height of the fracture surface outside the fractal regime. Here the distance r in the xy plane is measured from the centre of the crack.

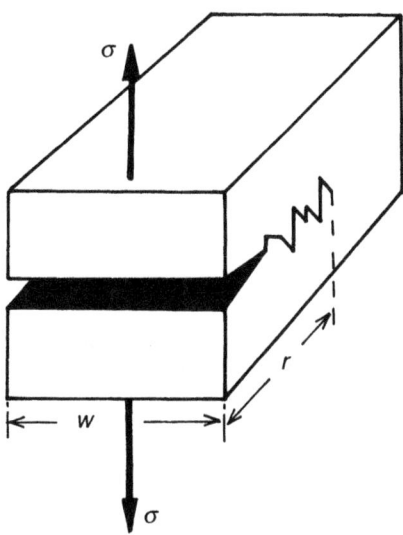

FIG. 3.7. A solid under stress σ (in mode I loading), containing a rough crack surface.

The equations (3.8) define a fractal crack surface such that its height z of the crack-tip grows with the distance r (in the xy plane) from the origin, following a power law, up to a length scale of the order of the (fractal) surface correlation length ξ, after which z saturates to a constant ($z_{\max}$). The power law growth of z for $r \ll \xi$ has clearly the self-affine property (3.6). The Griffith estimate for the fracture-toughness can then be made as follows (Mosolov 1993, Bouchaud and Bouchaud 1994): If a denotes the lattice constant, then integrating the elastic energy density $\sigma^2/2Y$, one gets

$$E_\text{l} \sim \left(\frac{\Lambda'^2 w}{2Y}\right) \int_a^l r^{-2\alpha} r \, dr,$$

or,

$$\sim \left[\frac{\Lambda'^2 w}{4(1-\alpha)Y}\right] l^{2(1-\alpha)}, \tag{3.9}$$

for the total elastic energy. For the crack surface height z, growing with crack-tip position r in the xy plane as $\sqrt{(dr^2 + dz^2)}$, one gets for the surface energy

$$E_\text{s} \sim 2\Gamma w \int_a^l \left[1 + \left(\frac{dz}{dr}\right)^2\right]^{1/2} dr, \tag{3.10}$$

where w denotes the width of the crack (plate) and Γ denotes the surface energy density. This can now be calculated in various situations:

$$E_s \sim 2\Gamma w\xi \left(\frac{z_{max}}{\xi}\right)\left(\frac{l}{\xi}\right)^\zeta \quad \text{for} \quad \frac{dz}{dr} \cong \frac{z}{r} \gg 1 \quad \text{and} \quad l \ll \xi, \quad (3.10a)$$

$$\sim 2\Gamma wl \quad \text{for} \quad \frac{dz}{dr} \cong 1 \quad \text{and} \quad l \ll \xi, \quad (3.10b)$$

and

$$\sim 2\Gamma w\left(\frac{z_{max}}{\xi}\right)l \quad \text{for} \quad \frac{dz}{dr} \simeq 0 \quad \text{and} \quad l > \xi. \quad (3.10c)$$

It is clear that both for the 'flat' surface case (with $l \gg \xi$ as in (3.10c), when the crack length is much higher than the correlation length of the fractal crack), and 'shallow' surface case (with $z_{max} \ll \xi$) and $dz/dr \cong 1$ as in (3.10b)), equating E_s with E_l one gets

$$\left[\frac{\Lambda'^2}{4(1-\alpha)Y}\right] l^{2(1-\alpha)} = 2\Gamma l,$$

giving $\alpha = 1/2$ and $\Lambda' = \Lambda = \sqrt{Y\Gamma}$. The variation of the strength of the solid with the crack length in such cases therefore is given by $\sigma(l) \sim \Lambda l^{-1/2}$, as in the Griffith case with smooth surfaces (3.5). For the 'rough' surface case (3.10a), equating E_s with E_l, one gets $\sigma_f \sim \Lambda' l^{-\alpha}$ with with $\Lambda' = \Lambda\sqrt{1-\alpha} = \Lambda\sqrt{\zeta/2}$. This again reduces to the usual result $\alpha = (2-\zeta)/2 = 1/2$ for the flat or nonrough surface case ($\zeta = 1$).

3.3.2 Experimental observations

As mentioned before, the crack surfaces in most of the natural (weakly disordered) solids are found to have a rough self-affine geometry with a universal value for the roughness exponent $\zeta \cong 0.80 - 0.85$, giving $\alpha \geq 1/2$ ($\cong 0.6$) (Bouchaud et al. 1993a,b, Roux 1994, Larralde and Ball 1995). In such cases of rough surfaces, one also gets the effective fracture toughness $\Lambda' \sim \Lambda\sqrt{\zeta/2}$ increasing with the increasing roughness of the surface. This square-root increase of the fracture toughness with roughness of the crack surface is also in good agreement with experimental observations (Mecholsky et al. 1988, 1989).

Maloy et al. (1992) have recently measured the fracture surface roughness exponent for different brittle materials, like random alloys or porcelains or graphites, by recording the heights at different positions (as a function of the position) along a one-dimensional cut of the fractured surface. Six different materials were studied. The fractured surfaces were traced, using essentially the gramophone pickup principle. A mirror was attached to a slightly weighted needle tracking along the surface, and a laser beam was reflected from the mirror. Using a position-sensitive photo-detector, which received the reflected beam, the heights $z(x)$ were recorded. The analysis (fit to eqn (3.6)) suggested a somewhat larger value for the roughness exponent: $\zeta \simeq 0.87$ for all the materials and (fractured) surfaces considered.

Kertesz (1992) and Kertesz et al.(1993) studied the morphology and the statistical properties of the self-affine fractal tear lines in a variety of papers under tensile stress. In sheets of size 30 cm × 45 cm, inserting a small notch of length 1 cm perpendicular to the stress and increasing the strain at the rate of 2 mm per minute using a tensile testing machine, the tearing morphology of the papers were studied with wide variations in the physical properties of the samples (variations of two or three orders of magnitude in paper quality, strength, density or toughness etc.). Statistical averages over about 150 samples gave the value of the roughness exponent ζ in the range 0.63-0.72. This roughness exponent corresponds very well with the exact result (Kardar et al.1986) for the surface roughness exponent (see e.g. Barabasi and Stanley 1994) or the transverse size exponent (Halpin-Healy and Zhang 1995) for directed polymers (or self-avoiding walks) in a disordered environment (lattice) in two dimensions.

This fractured surface may be modelled as a random surface passing through the cutting bonds, so that the surface has the least total energy. Obviously, the surface can not have any overhang: the height $z(x,y)$ can not be multi-valued at any point (x,y). This random surface problem is therefore analogous to that of a directed polymer in a random environment, in two dimensions, as mentioned before.

It may be mentioned here that all these results for the roughness exponent of fractured surfaces (or lines) are for quasi-static growth of fracture. As will be discussed later (in Section 3.7 on fracture dynamics), for rapid growth of fracture, the crack-tip instability (tip splitting and oscillations etc.), developed due to the nonlinear dynamics of the crack-tip (Langer 1993), gives rise to a crossover and a different roughness exponent for the fractured surfaces produced during the high velocity propagation of cracks (Abraham et al. 1994, Nakano et al. 1995, Abraham 1996).

3.4 Fracture strength of strongly disordered solids: fracture exponents near percolation threshold

In a disordered solid, as modelled by the percolation model (discussed in Section 1.2) with intact bond concentration p on a lattice, there occur various vacancy clusters (voids), or pre-existing cracks, of different sizes and shapes. The typical crack size in such a percolating solid being of the size of the correlation length (see Section 1.2.1) ξ, the application of the Griffith fracture criterion gives (Chakrabarti 1988, Ray and Chakrabarti 1985a,b) from equation (3.4)

$$\left(\frac{\sigma^2}{2Y}\right)\xi^d \cong \Gamma\xi^{d-1} \cong \xi^{d_B}, \tag{3.11}$$

where $\xi \cong |\Delta p|^{-\nu}$, $Y \cong (\Delta p)^{T_e}$, $\Delta p \equiv (p - p_c)/p_c$, p_c being the percolation threshold, and d_B is the backbone dimension (see Section 1.2.1). This assumes that the entire backbone of the percolating cluster forms the surfaces and shares (equally) the surface energy. This gives

$$\sigma_f \cong (\Delta p)^{T_f}; \quad T_f = \frac{1}{2}[T_e + (d - d_B)\nu], \tag{3.12}$$

for the variation of the typical fracture strength σ_f of the percolating solid with the bonding probability p near the percolation threshold p_c, for a fixed size of the sample. The brittle fracture exponent T_f is thus expressed in terms of (linear) elastic exponent T_e and the other exponents (and dimensions) representing the percolation cluster structure (statistics). Using the values of the exponent ν and backbone dimension d_B from Tables 1.2 and 1.3, one gets

$$T_f \simeq T_e/2 + 0.27, \quad (2D) \tag{3.12a}$$

and

$$T_f \simeq T_e/2 + 0.57, \quad (3D) \tag{3.12b}$$

where T_e is the elasticity exponent. As discussed in Section 1.2.1(f), $T_e \simeq 3.96$ and 3.75 in $d = 2$ and 3, which gives $T_f \simeq 2.2$ and 2.4 in the respective dimensions.

In the following, we give a more detailed analysis for estimating this fracture exponent T_f in various specific cases. There we also discuss various experimental investigations for determining the values of this fracture exponent.

3.4.1 *Estimates for the fracture exponent and comparisons with experiment*

(a) Discrete lattice system: theoretical estimates and experimental results

Using the node-link-blob model (see Section 1.2.1(d)) for the percolation cluster, one can in fact easily derive a rigorous bound of the fracture exponent T_f (Chakrabarti 1988, Ray and Chakrabarti 1988). This derivation is given below. One can see there that the above estimate for T_f in (3.12) turns out to be its lower bound.

As discussed in Section 1.2.1, for such systems near the percolation threshold p_c, the nearest-neighbour occupied bonds (or sites) form a statistically defined 'super-lattice', made of tortuous 'link-bonds' (of 'chemical length' L_c) crossing at nodes separated by an average distance ξ, the percolation correlation length (see Fig. 1.3 of Chapter 1). The external stress

σ is shared by $\xi^{-(d-1)}$ number of parallel links, giving the stress per link $\sigma_L \cong \sigma\xi^{(d-1)}$. The total strain

$$\epsilon = \sigma/Y \cong \sigma\xi^{(T_e/\nu)}$$

is shared by ξ^{-1} number of link-bonds, giving the strain per link-bond

$$\epsilon_L \cong \sigma\xi^{(1+T_e/\nu)}.$$

This gives the elastic energy per link-bond

$$E_L \cong \sigma_L\epsilon_L \cong \sigma^2\xi^{(d+T_e/\nu)}. \qquad (3.13)$$

If this entire elastic energy of the link is shared equally by all the $M = \xi^{d_B}$ (elemental) bonds of the link-bond (including the dangling ends), then the energy per bond E_L/M is of the order of $\sigma^2\xi^{[(d-d_B)+T_e/\nu]}$. Because of the assumption of equal sharing by all the M bonds of the link, this gives an underestimate of the strain-energy per bond, as many of the elemental bonds do not support any stress or support very small stress because of the multiply connected loop structure of the percolating lattice. If one now assumes a fixed energy threshold for each bond, the fracture will occur beyond $\sigma = \sigma_f$ for which E_L/M goes beyond the threshold value:

$$\sigma_f \cong (\Delta p)^{T_f}; \text{ with } T_f \geq \frac{1}{2}[T_e + (d-d_B)\nu].$$

If all the strain energy is supported by the $L_c \sim \xi^{1/\nu}$ number of singly connected bonds only (see Section 1.2.1), then one gets an overestimate of the elastic energy per bond, which comes out to of the order of $E_L/L_c \sim \sigma^2\xi^{(d+T_e/\nu-1/\nu)}$. When one equates this energy with the threshold energy of the bond, one gets the lower bound:

$$T_f \leq \frac{1}{2}[T_e + d\nu - 1].$$

Combining these, one thus gets the rigorous bounds for T_f (Ray and Chakrabarti 1988, Chakrabarti 1988)

$$\frac{1}{2}[T_e + d\nu - 1] \geq T_f \geq \frac{1}{2}[T_e + (d-d_B)\nu]. \qquad (3.14)$$

For $d \geq 6$, the links of the super-lattice network are indeed made up of singly connected bonds (see Section 1.2.1), and therefore both the bounds become equal and equalities in (3.14) become exact. The above bounds are valid in general for any elastic network, provided the exponents like T_e, ν, d_B, etc. are appropriate for the network under consideration. For

Table 3.1 *Theoretical estimates for the fracture exponent T_f*

Dimension (d)	Lattice percolation	Continuum percolation
2	$\frac{1}{2}(T_e + 0.4\nu) \leq T_f \leq \frac{1}{2}(T_e + 2\nu - 1)$ $2.2 \leq T_f \leq 2.8$	$\frac{1}{2}(T_e + 2\nu + 3/2)$ 4.1
3	$\frac{1}{2}(T_e + 1.3\nu) \leq T_f \leq \frac{1}{2}(T_e + 3\nu - 1)$ $2.4 \leq T_f \leq 2.7$	$\frac{1}{2}(T_e + 3\nu + 5/2)$ 4.5

systems with both bond-stretching and bond-bending forces, $\nu = 1/2, d_B = 2$ and $T_e = 4$ for $d \geq 6$, giving $T_f = 3$ for $d \geq 6$; and this is an exact result (Bergman 1986, Chakrabarti 1988). Using the results for the elastic exponents d_B, ν and T_e (see Section 1.2.1), one gets (Chakrabarti 1988) $2.2 \leq T_f \leq 2.8$ and $2.4 \leq T_f \leq 2.7$ in $d = 2$ and 3 respectively (see Table 3.1). The corresponding estimates for the bounds of T_f for central force networks are not possible because of the absence of detailed knowledge about the exponents like ν etc. for such neworks (see Section 1.2.1(f) and Table 1.4). In any case, all these results are for a fixed sample size L, which is large but finite.

It may be noted here that the above lower bound for the exponent T_f in (3.14) is actually the estimate (3.12) of the same exponent, obtained using the Griffith law with the typical crack size scaling as ξ and the surface energy scaling as ξ^{d_B}. Although the details of the calculations turn out to be intrinsically similar, the above-mentioned coincidence in the estimates of T_f using two different methods is quite intriguing. It may also be mentioned here that assuming the total strain of the network to be related to the strain of a single cell of the node-link-blob model, and assuming that to be determined by the number of singly connected bonds in a link, Bergman (1986) (see also Benguigui et al. 1987) obtained the bounds $T_e - \nu d_{\min} \leq T_f \leq T_e - 1$, where $d_{\min}$ is the fractal dimension of the shortest path (see Section 1.2.1). Here, the strain per bond is equated to a threshold value rather than the elastic energy per bond, as was used in the previous derivation. Employing the values of $d_{\min}$ ($\simeq 1.13$, 1.34 and 2 in $d = 2, 3$ and 6 respectively from Table 1.3) and ν, one gets here $2.45 \leq T_f \leq 2.96$ in $d = 2$, $2.58 \leq T_f \leq 2.76$ in $d = 3$ and $T_f = 3$ for $d \geq 6$ when both the bounds coincide.

Experimental verifications These estimates for T_f are in agreement with the experimental results for the fracture stress σ_f in randomly perforated (on lattice) thin sheets of aluminium and copper (Benguigui et al. 1987, Sieradzki and Li 1986), where the results are comparable to the two-dimensional theoretical results.

Benguigui et al. (1987) performed the fracture experiments for such

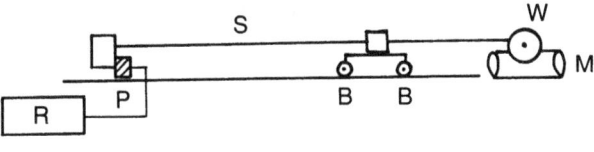

P - pressure sensor
R - recorder
S - sample
B - ball bearing
M - motor
W - wheel

(a)

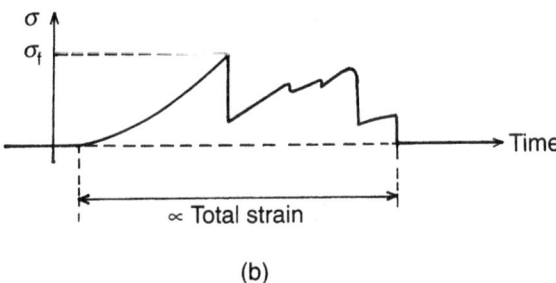

(b)

FIG. 3.8. (a) The experimental set-up for investigating fracture of thin metal foils containing random network of punched holes. (b) A typical record of the stress with time during the fracture process. Each discontinuity corresponds to the breaking of a bond. σ_f denotes the stress at the first breaking, and is usually the largest stress compared to those for the successive breakings. The time for total breaking gives the total strain before breaking.

metallic sheets using two different modes. In first, called the strain mode, the sheet is elongated very slowly and the stress is measured. In the second (stress mode), the stress is increased until fracture and the elongation is measured. The samples are prepared either by punching random holes of diameter slightly greater than the unit cell of the square lattice drawn on thin copper or aluminium foils (used in the first mode), or by cutting the interhole bonds at random when the whole size is smaller than the lattice unit cell (used in second mode).

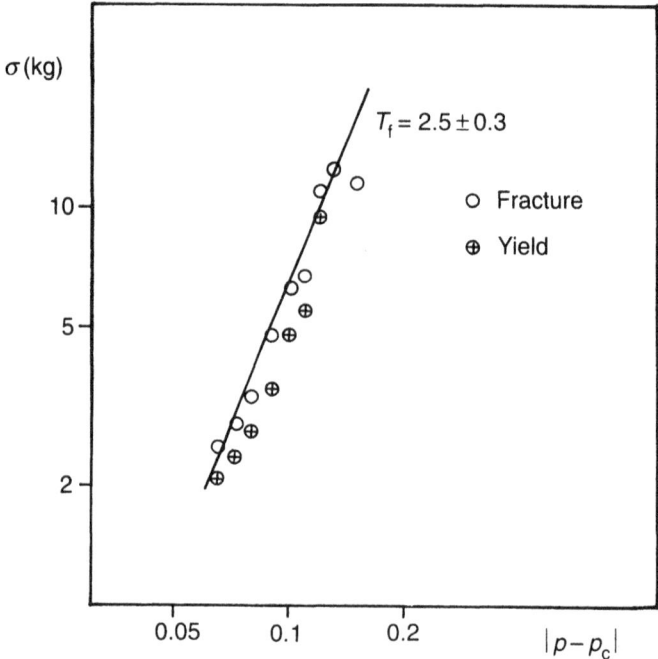

FIG. 3.9. Fracture load (o) and the yield load (⊕) versus $(p - p_c)$ in a log-log plot (Benguigui et al. 1987).

In the first (strain) mode, the stresses necessary to break the different bonds are measured. It is found that the stress necessary to break the first bond is always larger than those necessary for the breaking of the successive bonds (see Fig. 3.8). In accordance with this result, the fracture occurs by a cascade effect in this mode. The fracture stress is defined as the stress necessary to break the first bond in the strain mode, and by the stress which gives the fracure of the sample in the stress mode. At the same time, the total elongation of the sample until complete fracture is also recorded in the two modes.

Two exponents were determined in these experiments: that for the fracture stress, which goes to zero at p_c, and that for the elongation of the sample, which seems to diverge at p_c. There is also complete agreement between the results of the two kinds of measurements. The fracture exponent T_f is found to be equal to 2.5 ± 0.3, well within the bounds (3.14) already calculated above (see Fig. 3.9). The elongation before fracture is found to grow with $(p - p_c)$ having an exponent value 1.4 ± 0.2.

The fact that the fracture is ductile, and not brittle, may give rise to the suspicion that the agreement with the theory is only accidental. In fact, this is not the case. We shall discuss this point later (see Section 3.4.2). For the time, it can only be mentioned that the yield stress, for which the deviation from from linear elasticity starts, decreases to zero as p goes to p_c, with an exponent value slightly less than that for the fracture stress (see Fig. 3.9).

Sieradzki and Li (1986) studied the entire stress-strain curve of such samples until fracture. They however obtained $T_f \cong 1.7 \pm 0.1$, which is much below that obtained by Benguigui et al. (1987), and also below the minimum theoretical bound we obtained earlier. However, this low value of T_f is most likely because of analysing the data for regions far away from the critical point.

Results of computer simulations Attempts have been made to check this fracture growth behaviour and fracture exponents in molecular dynamic simulations (Ray and Chakrabarti 1985b, Chakrabarti et al. 1986). Here, the simulation program is the same as described in Section 3.2.3, where the initially (intact) bonds, between the occupied atoms, transmit interactions governed by Lennard-Jones potential with a cut-off strain or interatomic distance. The external force or stress is again applied at the surfaces. The initial lattice has a certain fraction of intact bonds (between the randomly placed atoms occupied with concentration p), while the rest of the bonds are inactive. This is simply done using a Monte Carlo program. This random removal of bonds or sites gives the initial void clusters or microcracks, having different shapes and sizes.

The molecular dynamic simulation starts at this point, and the appropriate displacements (calculated for suitably chosen small time interval δt) of the atoms are obtained from the instantaneous calculation of the local forces governed by the Lennard-Jones potential corresponding to the relative positions of the neighbouring atoms, connected by intact bonds. With the successive (time) iterations, the lattice starts getting deformed and the stresses concentrate around the sharp notches of the voids or microcracks. After the network relaxes, when the resultant force on any nodal point of the network decreases below an arbitrarily chosen small value, the external stress is increased by a small amount. Eventually, with the increase in the external stress, the interatomic separation at the point of highest stress concentration exceeds the threshold value, and rupture occurs there (see Fig. 3.10). This stress is taken as the fracture nucleation stress of the specimen. In such networks with random initial microcracks, the occurrence of any such rupture or nucleation of the fracture does not necessarily imply its propagation (as in the nonrandom case discussed in Section 3.2.3), and the fracture propagation may get arrested if sufficient surface energy can not be provided at the fracture tip.

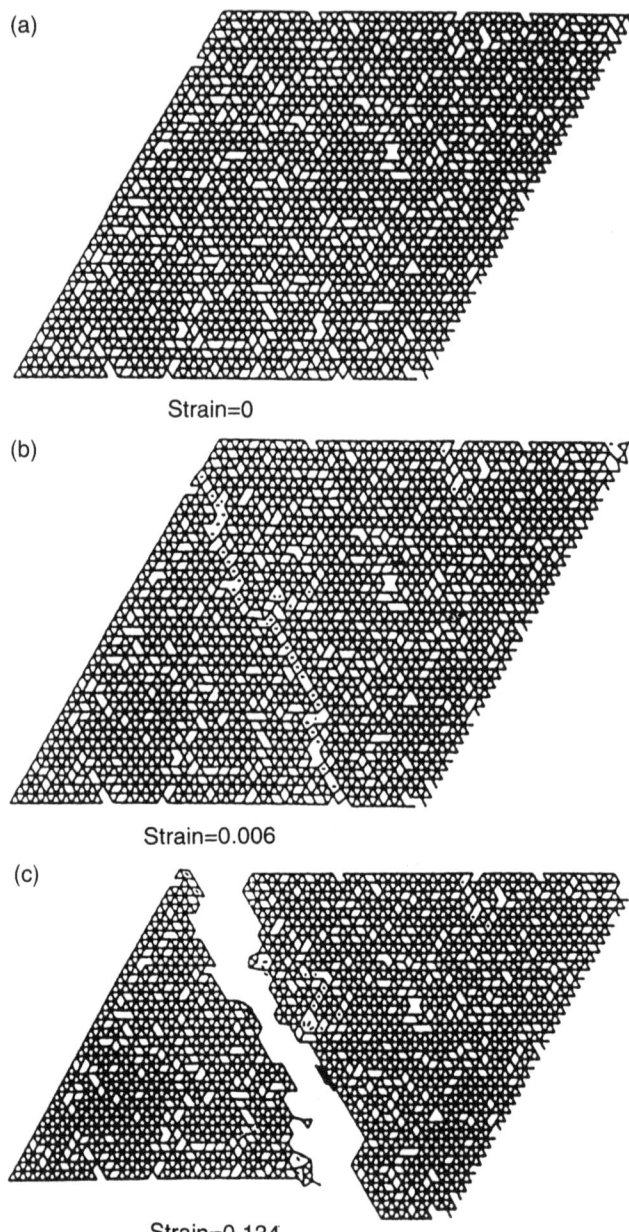

FIG. 3.10. (a) A random lattice configuration with bond (spring) concentration $p = 0.90$, before the application of stress. (b) The configuration after the rigidity failure. (c) The configuration after complete fracture. Computer simulation results from Beale and Srolovitz (1988).

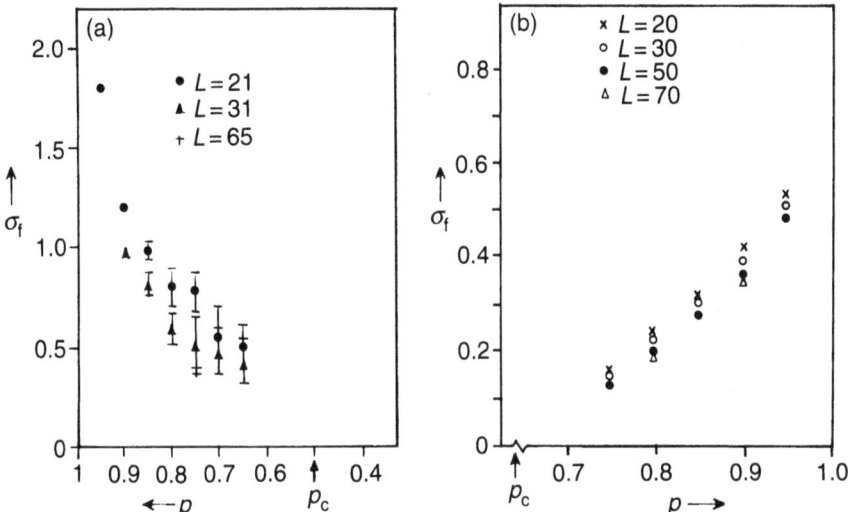

FIG. 3.11. Molecular dynamic simulation results for the average fracture stress σ_f for various disorder concenrations on triangular lattices. (a) For site dilute Lennard-Jones system (Chakrabarti et al. 1986), and (b) for bond dilute spring network (Beale and Srolovitz 1988).

Figure 3.11(a) shows the variation of this (configurationally averaged) minimum fracture stress with the initial lattice dilution concentration $1-p$, as obtained in the molecular dynamic simulation for a triangular lattice Lennard-Jones system with site dilution (Chakrabarti et al. 1986). The average fracture stress is clearly seen to decrease continuously with the increasing (initial) lattice dilution. However, such molecular dynamic studies have not been made accurate enough to investigate the critical properties and obtain the fracture exponent T_f. Although the molecular dynamic study of fracture is the most straightforward, pure, yet feasible (in principle) numerical method, its application is strongly limited (in practice) by the heavy consumption of computer time for such randomly disordered systems.

Alternative numerical methods exist, where one minimises the total elastic energy E of the network with the lattice displacement vector $\mathbf{u}_i$: $\partial E/\partial \mathbf{u}_i = 0$ for all the internal modes of the network. This gives dN simultaneous equations for the displacement vector $\mathbf{u}_i$ for the N internal modes of the network. The resulting set of linear equations is solved numerically using, for example, the Jacobi conjugate gradient method (see e.g. Beale and Srolovitz 1988, Sahimi and Arbabi 1993). Figure 3.11(b) shows the variation of the average fracture (nucleation) stress σ_f, obtained using the above method for bond dilute or random spring triangular lattice networks (Beale and Srolovitz 1988). This result is comparable with the molecular dynamic simulation (Chakrabarti et al. 1986), shown in Fig. 3.11(a).

Accurate numerical studies, using this self-consistent strain calculation method for various elastic networks, with central and bond-bending forces, by Sahimi and Arbabi (1993) gave $T_f \cong 2.42 \pm 0.14$ and $\cong 2.64 \pm 0.3$ in $d = 2$ and 3 respectively. These estimates again compare well with the experimental results of Benguigui et al. (1987) in $d = 2$, and fall within the above-mentioned theoretical bounds (given by (3.14)).

(b) Continuum systems: theoretical estimates and experimental results

In the discrete lattice model, discussed above, each bond is identical, having identical threshold values for its failure. In the laboratory simulation experiments (discussed in the previous section) on metal foils to model such systems, holes of fixed size are punched on lattice sites and the bonds between these hole sites are cut randomly. If, however, the holes are punched at arbitrary points (unlike at the lattice sites as discussed before), one gets a 'Swiss-cheese' model of continuum percolation. For linear responses like the elastic modulus Y or the conductivity Σ of such continuum disordered systems, there are considerable differences (Halperin et al. 1985) and the corresponding exponent values for continuum percolation are higher compared to those of discrete lattice systems (see Section 1.2.1 (g)). We discuss here the corresponding difference (Chakrabarti et al. 1988) for the fracture exponent T_f. It is seen that the fracture exponent $\tilde{T}_f$ for continuum percolation is considerably higher than that T_f for lattice percolation: $\tilde{T}_f = T_f + (1 + x)/2$, where $x = 3/2$ and $5/2$ in $d = 2$ and 3 respectively.

In the node-link model of the 'Swiss-cheese' system (see Section 1.2.1 (g)), each link has a distribution $P(\delta)$ of channel width δ and hence of local strength, as indicated in Fig. 1.5 of Chapter 1. One expects here the minimum value of $\delta = \delta_{\min}$ to be of the order of L_c^{-1}, where $L_c \sim (\Delta p)^{-1} \sim \xi^{1/\nu}$ is the chemical length of the singly connected link of the percolating network, and $P(\delta)$ to be finite for $\delta \to \delta_{\min}$.

The strain energy E_L per link-element is again given here by (3.13), where T_e is replaced by $\tilde{T}_e = T_e + x$, $x = 3/2$ and $5/2$ for $d = 2$ and 3 respectively, for the continuum systems (see Section 1.2.1(g)). However, since the bonds in the links are not uniform in strength, there will be a distribution of $E(\delta)$. This gives the strain energy for a bond with channel width δ as

$$E_L = \int_{\delta_{\min}}^{\infty} E(\delta) P(\delta) d\delta. \quad (3.15)$$

If $E(\delta) \sim \sigma^2 \delta^{-a}$, this gives $E_L \sim \sigma^2 L_c^a \sim \sigma^2 \xi^{a/\nu}$. Comparing now with E_L in (3.13), we get $a = d\nu + \tilde{T}_e$. The elastic energy for the weakest bond (with $\delta = \delta_{\min} \sim 1/L_c$) is then of the order of $\sigma^2 L_c^a$, giving the limiting applicable stress $\sigma_f \sim L_c^{-a/2} \sim (\Delta p)^{\tilde{T}_f}$, where

$$\tilde{T}_f = \frac{1}{2}(\tilde{T}_e + d\nu) = \frac{1}{2}(T_e + d\nu + x). \qquad (3.16)$$

It may be noted that since the weakest link breaks here and the strain energies of other stronger bonds are negligible, the multiple connections in the links are not considered, as they effectively belong to stronger bonds. In $d = 2$, one then gets $\tilde{T}_f \cong 4.06$ (using $T_e \cong 3.96$ and $\nu = 4/3$) and in $d = 3$, $\tilde{T}_f \cong 4.45$ (using $T_e \cong 3.75$ and $\nu \cong 0.89$) for the fracture exponent values in continuum (Chakrabarti et al. 1988). In a slightly different way, using the same limiting strain energy concept for fracture in the 'Swiss-cheese' model, Sornette (1987) obtained $\tilde{T}_f \cong 3.92$ and 4.39 in $d = 2$ and 3 respectively. These estimates are indeed very close. This result for $d = 2$ seems also to compare with the experimental observation of Benguigui ($\tilde{T}_f \cong 4.0 \pm 1.0$) for fracture stress in randomly punched metal foils (unpublished; see Chakrabarti 1988).

All the above theoretical estimates for the lattice and continuum values of the fracture exponent T_f are given in Table 3.1.

3.4.2 Brittleness and plastic yield

As we have mentioned in the introduction (Chapter 1), and also at the beginning of this chapter, all these scaling relations are derived assuming brittleness (linear stress-strain relationship up to the breaking point) of the samples near p_c. In the experiments (Benguigui et al. 1987, Sieradzki and Li 1986), although the foils used (aluminium or copper) are not at all brittle to start with, the results for the fracture strength not only compare well with the theoretical estimates obtained assuming microscopic brittleness, they indeed show almost perfect brittleness near p_c. Benguigui et al. (1987) measured the fracture strain ϵ_f growing near p_c as $\sim (\Delta p)^{-T_s}$, where $T_s = 1.4 \pm 0.2$, compared to the fracture stress ($\sigma_f \sim (\Delta p)^{T_f}$) exponent $T_f = 2.5 \pm 0.4$ in $d = 2$. If we assume stress-strain linearity up to the breaking point, then $\sigma_f = Y\epsilon_f$, giving the value for the exponent T_e for the elasticity modulus Y as $T_e = T_s + T_f \cong 3.9$, which is precisely the best theoretical estimate (Zabolitzki et al. 1986) and close to the measured value (Benguigui et al. 1987) of the elasticity exponent.

In fact, if we denote by σ_y ($\leq \sigma_f$) the stress at which the irreversible plastic deformation occurs and if $\sigma_y \sim (\Delta p)^{T_y}$, then one can estimate T_y in the lattice cases, using the following picture due to Bergman (1986). As one approaches p_c, the singly connected bonds in the links are strained more than other bonds (in the blobs etc.). Plasticity occurs when some of these bonds are strained (irreversibly) beyond their linear range. Assuming now the other bonds in the blobs etc., which are very weakly strained because of their favourable positions, to be completely rigid, the problem of plasticity (finding σ_y) becomes identical to the problem of fracture (finding σ_f) in a random rigid-bond and elastic-bond network. Thus from equation (3.14)

$$\frac{1}{2}[T'_e + d\nu - 1] \geq T_y \geq \frac{1}{2}[T'_e + (d - d_B)\nu]. \qquad (3.17)$$

Here T'_e is the elasticity exponent of a random rigid-bond network, and T'_e appears (Bergman 1986) to be identical to the conductivity exponent t_c (see Section 1.2.1(c)). Since $T_e > t_c$, we get $T'_e < T_e$, suggesting $T_y < T_f$: for example, $1.5 \geq T_y \geq 0.9$ and $2.8 \geq T_f \geq 2.3$ in $d = 2$. This also suggests that the relative ratio $r = (\sigma_f - \sigma_y)/\sigma_f$ of the nonlinear range (plastic yield) response shrinks exponentially as $\exp[-(\Delta p)^{-T_r}]$ with $T_r = T_f - T_y$. As mentioned previously, this shrinking of the plastic region is also supported by the observed (Benguigui et al. 1987) perfectly linear relationship of fracture stress and strain, although the metals used are highly ductile to start with.

3.5 Fracture strength distribution

The fracture stress exponents discussed above represent the singularities in the variations of the strength with disorder for the average fracture stress with fixed, very large, but finite sample size. However, as is well known and established by now (Jayatilaka 1979, Duxbury 1990, Leath and Duxbury 1994), the breakdown strength distribution of disordered solids is not quite self-averaging (giving different most-probable and average values), as the statistics is governed by the extreme events. As already discussed in Section 1.2.2(b) on extreme statistics, and shown for the electrical breakdown cases (Section 2.2.2(c)), this non-self-averaging property of the breakdown strength distribution gives rise to nontrivial sample size dependence of the average fracture stress or breakdown strength, in the sense that the breakdown strength vanishes in the truly infinite size limit for any nonvanishing concentration of disorder. This happens due to the fact that the strength of a solid is determined by that of the weakest point in the entire sample, and not by any kind of averages over such weak points. We now study the distribution of fracture strengths of such random solids.

3.5.1 Extreme statistics and strength distribution

In what follows, we will show that, similar to that for the electrical breakdown (Section 2.2.2(c)), the cumulative probability of failure or fracture

$F(\sigma)$ of a sample of linear size L containing random defects can have the following two forms:

$$F(\sigma) \sim 1 - \exp\left(-\frac{L^d \sigma^m}{(\Lambda(p))^m}\right), \qquad (3.18a)$$

or,

$$F(\sigma) \sim 1 - \exp\left[-L^d \exp\left(-\frac{c(\Delta p)^{2T_f}}{\sigma^{1/\psi}}\right)\right]. \qquad (3.18b)$$

Here $\Lambda(p)$ is a function of the disorder concentration p, the power m is called the Weibull modulus, T_f is the fracture exponent and c and ψ are constants. The first distribution (3.18a) is called the Weibull distribution, while the other one is called the Gumbel distribution.

The average fracture stress σ_f can then be obtained setting $F(\sigma)$ finite: $F(\sigma_f) \simeq 1/2$. This gives

$$\sigma_f \sim \frac{\Lambda(p)}{L^{d/m}}, \qquad (3.19a)$$

or,

$$\sigma_f \sim \frac{(\Delta p)^{2\psi T_f}}{(\ln L)^\psi}, \qquad (3.19b)$$

for the Weibull and Gumbel distributions respectively. Both these forms clearly show that the average fracture stress vanishes ($\sigma_f \to 0$), although very slowly (logarithmically or with extremely small power), for any disordered solid in the truly macroscopic size limit ($L \to \infty$). This is because of the finite probability of getting an infinite crack (or a dangerously weak point) due to statistical fluctuations, and fracture being a catastrophic phenomenon, the weakest point determines the strength of the entire solid.

Let us assume that a solid of linear size L contains n number of cracks, each with failure probability $f_i(\sigma), i = 1, 2, .., n$, under an applied stress σ. We further assume that the stress-released regions of each of the cracks are separate and do not overlap. If we denote the cumulative failure probability of the entire sample under stress σ by $F(\sigma)$, then, as discussed in Section 1.2.2(b), we get

$$1 - F(\sigma) \cong \exp[-L^d g(\sigma)], \qquad (3.20)$$

where $g(\sigma)$ denotes the density of cracks weaker than stress σ (or the density of cracks which will start propagating at and above the stress σ). This is because the sample survives if each individual crack within the sample survives.

We will give next two specific and limiting forms of this distribution, indicating their derivations for percolating solids. It should be noted, however, that these forms are not specific to percolation models only; they can be applied to much wider classes of disorder.

(a) Weibull distribution

This distribution appears whenever $g(\sigma)$ is given by a power law in σ, coming from the power law variation of the density of linear cracks $g(l)$ with their length l. In the random percolation model considered here, this does not normally occur (except at the percolation threshold $p = p_c$). However, for various correlated disorder models, applicable to realistic disorders in rocks, composite materials, etc., one can have such power law distribution for clusters, which may give rise to a Weibull distribution for their fracture strength. We will discuss such cases later, and concentrate on the random percolation model in this section.

For small bond dilution concentration, the probability $g(l)$ of a linear crack of length l is given by $g(l) \cong p^2(1-p)^l$. This gives (Ray and Chakrabarti 1985a)

$$g(\sigma) = \int_{l(\sigma)}^{\infty} g(l) dl \cong \left[\frac{p^2}{\ln(1-p)}\right](1-p)^{(\Lambda/\sigma)^2}, \qquad (3.21)$$

as the minimum length of the crack growing under stress σ is given by $l(\sigma) = (\Lambda/\sigma)^2$ from the Griffith law (3.5). Although this $g(l)$ function, and hence $g(\sigma)$, is not any power function, rather an exponential function in l or σ, one can approximately represent them by power laws. Note that as σ varies from 0 to ∞, $(1-p)^{(\Lambda/\sigma)^2}$ varies from 0 to 1 ($p < 1$); and this suggests a possible representation (Ray and Chakrabarti 1985a, see also Zhang et al. 1990) $g(\sigma) \cong 1 - \exp[-c(\sigma/\Lambda)^m] \cong c(\sigma/\Lambda)^m$, with the fitting parameter (Weibull modulus) $m(p) > 0$. This gives the Weibull failure strength distribution (Ray and Chakrabarti 1985a, Chakrabarti 1988)

$$F(\sigma) \cong 1 - \exp\left[-cL^d\left(\frac{\sigma}{\Lambda}\right)^m\right],$$

giving

$$\sigma_f \cong \frac{\Lambda(p)}{L^{d/m}},$$

for the variation of the typical strength of the sample (corresponding to a nonzero value of $F(\sigma)$), as a function of sample size L.

(b) Gumbel distribution

For very large amounts of disorder, we can use the scaling theory for the variation of the typical vacancy cluster or void size with disorder. Very near the percolation threshold p_c, the probability of getting a vacancy cluster (crack) of linear size l will be given by (see Section 1.2.1)

$$g(l) \approx \exp(-l/\xi) \cong \exp\left(-\frac{\Lambda^2}{\sigma^2 \xi}\right), \qquad (3.22)$$

where the Griffith law (3.5) has again been used to relate the minimum length l of the cracks which become unstable under stress σ. Using again (3.5), one gets $\Lambda^2/\xi = Y\Gamma/\xi \cong \Delta p^{2T_f}$, where $T_f = [T_e + (d - d_B)\nu]/2$. This gives the Gumbel distribution $F(\sigma)$ for the strength of the entire sample (Chakrabarti 1994)

$$F(\sigma) \cong 1 - \exp\left[-L^d \exp\left(-\frac{(\Delta p)^{2T_f}}{\sigma^{1/\psi}}\right)\right].$$

This gives

$$\sigma_f \cong \frac{(\Delta p)^{T_f}}{(\ln L)^\psi},$$

for the variation of the typical strength of the solid as a function of the sample size L with $T_f = [T_e + (d - d_B)\nu]/2$ and $\psi = 1/2$. It may be noted here that this value of ψ ($= 1/2$) is obtained here using the Griffith law for the individual cracks within the solid, having smooth or non-rough crack surfaces. As discussed earlier (in Section 3.3), for rough crack surfaces, the value of ψ can be higher.

It may be mentioned here that this result is valid for continuum percolation as well, with the appropriate value ($= \tilde{T}_f$) for the exponent T_f.

3.5.2 Comparisons with computer simulational and experimental results

(a) Computer simulation results

The validity of these fracture strength distributions (3.18a) and (3.18b) has not been checked yet extensively in experiments, as the above two forms differ very little numerically and require very accurate data for the failure strength distribution $F(\sigma)$ for the analysis (see however the next subsection). Various accurate numerical simulation experiments have however been performed.

The fracture strength distribution of a two-dimensional triangular network of randomly diluted bond percolating systems with central force was obtained by both Beal and Srolovitz (1988) and Sahimi and Arbabi (1993), using various numerical techniques, including the conjugate gradiant method discussed in Section 3.4.1. The results for such central force systems showed that the Gumbel distribution (3.18b), with $\psi \simeq 1/2$, fits the data better than the Weibull distribution. This feature was found to be valid for the entire range of dilution concentration, up to the (central force) percolation threshold.

Sahimi and Arbabi (1993) also studied the fracture strength distribution of a two-dimensional (triangular lattice) randomly diluted network with both central and bond-bending forces (with Hamiltonian given by (1.11) in Section 1.2.1 (f)). The results showed that, although the Weibull distribution fits the data initially for small disorder (p near unity), the data fits the Gumbel distribution considerably better and much more accurately as disorder increases ($p \to p_c$). In fact, one can define a quantity A as

$$A = \ln\left[-\frac{1}{L^d}\ln(1 - F(\sigma))\right], \quad (3.23)$$

and plot A against $\ln(\sigma)$ or $1/\sigma$ to distinguish between the distributions. For the Weibull distribution

$$A \sim \ln \sigma, \quad (3.23a)$$

while

$$A \sim \frac{1}{\sigma^{1/\psi}}, \quad (3.23b)$$

for the Gumbel distribution. The results of Sahimi and Arbabi (1993) for network size $L = 60$ (with $\beta = \kappa_b/\kappa_c = 0.1$ in the Hamiltonian (1.11) and the bonds breaking beyond a fixed amount of stretching) gave quite good fit with the Gumbel distribution (with $\psi \simeq 1$). Figure 3.12 shows such fits for bond occupation probabilities $p = 0.9$ and 0.5 ($p_c \simeq 0.35$).

For the fracture strength distribution of superelastic networks (see Section 1.2.1(f)), containing bonds of finite strength, which break beyond a fixed amount of stretching, and infinitely rigid bonds, which do not break, Sahimi and Arbabi (1993) however observed the Weibull distribution to fit better than the Gumbel distribution.

(b) Experimental results

The mechanical strength of highly porous ceramics like cylindrical silica extrudates were studied by van den Born et al. (1991). They measured the failure strength distribution $F(\sigma)$ for (four) different series of samples produced under different conditions, resulting in different porosity and other porometric parameters. The failure strength distribution $F(\sigma)$ (for normalised constant sample volume) is shown in Fig. 3.13(a). The plot of A, as defined above in (3.23), against $\ln \sigma$ (in Fig. 3.13b) and $1/\sigma$ (in Fig. 3.13c) shows that the fit with the Gumbel distribution (3.18b), with $\psi \simeq 1$, is much better.

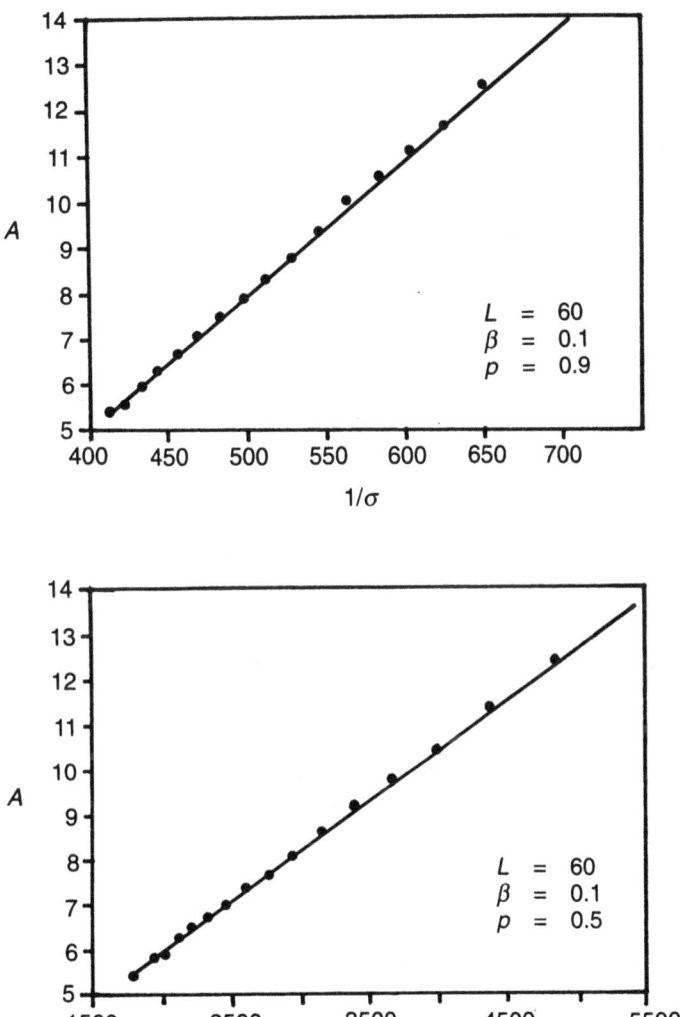

FIG. 3.12. Fit to Gumbel distribution for the computer simulation results of fracture strength for triangular network of springs with bond bending force ($\beta = 0.1$), with the linear size L of the network fixed ($L = 60$). Plot of A versus $1/\sigma^{1/\psi}$ with $\psi = 1$. (a) For $p = 0.9$ and (b) $p = 0.5$ (Sahimi and Arbabi 1993).

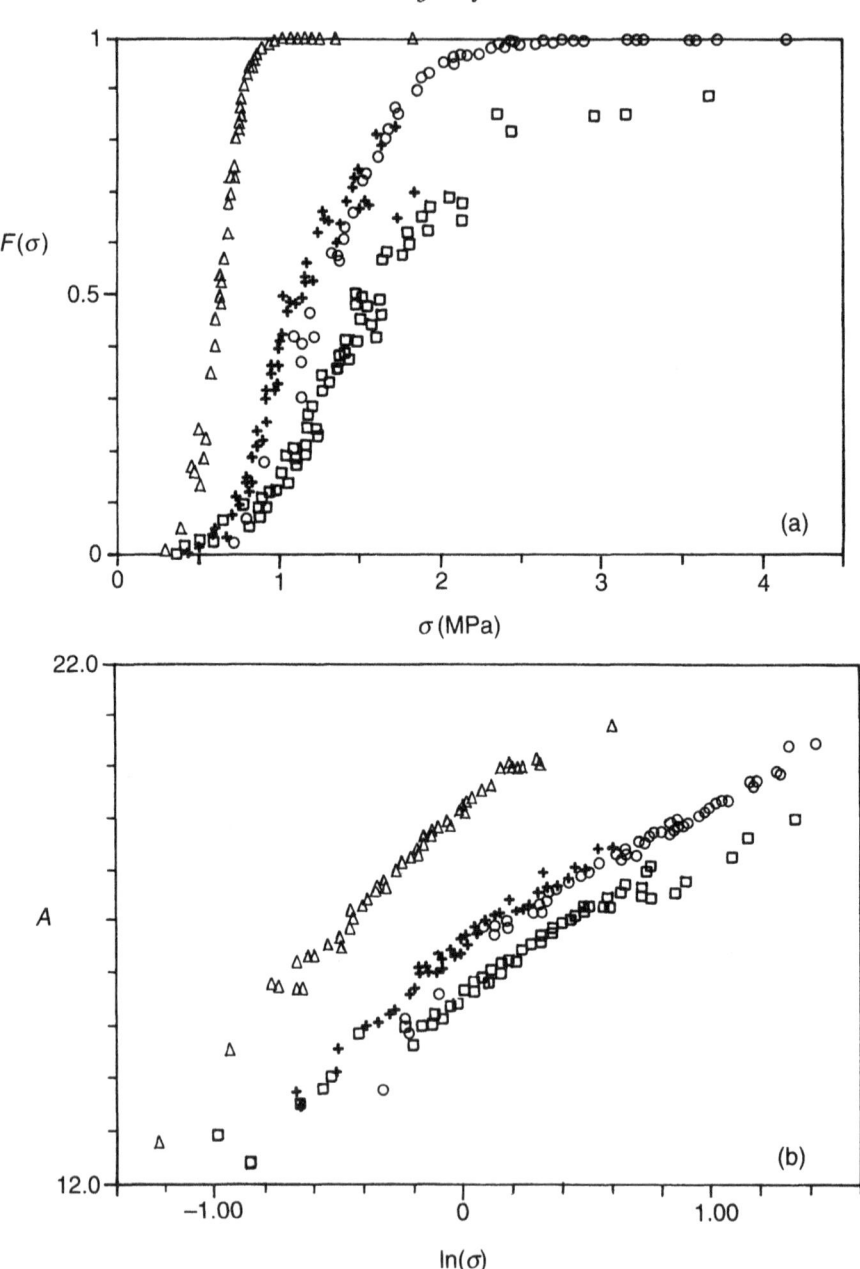

FIG. 3.13. Experimental data for the fracture strength distribution of porous silica extrudates. (a) Cumulative failure distribution $F(\sigma)$, normalised to a constant volume, (b) Weibull distribution fit (A versus $\ln \sigma$)

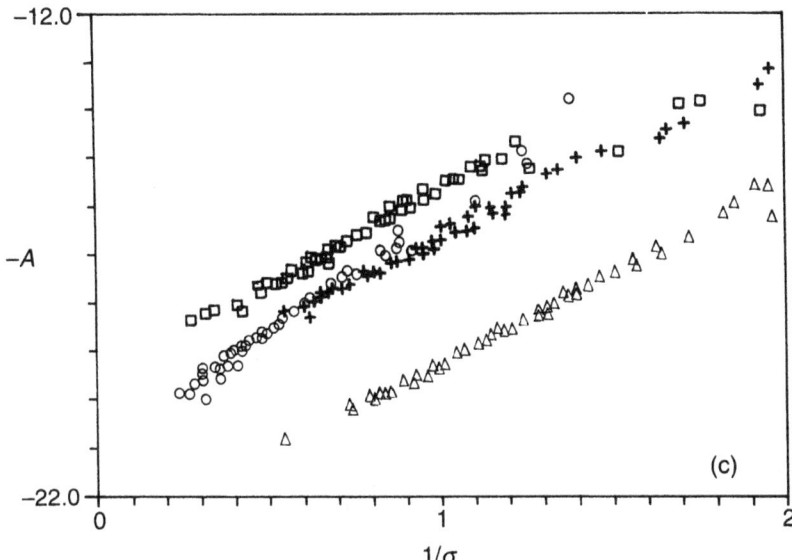

FIG. 3.13. (c) Gumbel distribution fit (A versus $1/\sigma$). Different symbols correspond to different series of samples (porosity etc.) (van den Born et al. 1991).

3.6 Fracture strength scaling in systems with random breaking thresholds

So far, we have considered systems with randomly broken and intact bonds (before the stress is applied and fracture occurs), with the intact bonds having a fixed breaking threshold (for example, breaks for strains beyond a fixed interatomic separation a_c). However, to model the fracture process in a large class of inhomogeneous mixtures, composites and granular materials, one needs to consider systems where the intact bonds may have a (continuous) strength distribution. This can be achieved in two kinds of model systems (Sahimi and Goddard 1986, Sahimi and Arbabi 1993): (i) with randomly distributed spring constant (κ), but with fixed limiting interatomic separation a_c, beyond which the bonds break where one may or may not start with a fraction of broken springs (for which $\kappa = 0$); (ii) for the other kind of models, one fixes again the spring constant at a value identical for all the springs but the critical or threshold strains a_c are assumed to be a randomly distributed quantity. The scaling behaviour of the fracture strength distribution of such models, in particular for the models of the second kind (ii), appear to be significantly different (see de Arcangelis 1990, Sahimi and Arbabi 1993) from those discussed earlier (for percolating solids).

3.6.1 Models with random spring constants

With uniform distribution of the spring constant (κ) in the interval 0 to 1, Sahimi and Goddard (1986) found using computer simulations (molecular dynamic) that unless a significantly large fraction of the springs is broken ($\kappa = 0$) initially, only a single crack is formed, which propagates throughout the system, eventually breaking the sample into two pieces. They observed that although some side branches are indeed developed occasionally, they remain arrested to short sizes, such that they do not contribute any significant statistical effect. Similar results were observed with a log-normal distribution of spring constants. With the distribution

$$P(\kappa) = (1-\alpha)\kappa^{-\alpha}; \quad 0 < \alpha < 1, \tag{3.24}$$

similar to that considered by Halperin et al. (1985) for their random void elastic network model (see Section 1.2.1(g)), Sahimi and Goddard (1986) observed in their molecular dynamic simulation studies of spring (central force) networks on a triangular lattice that for $\alpha \cong 0$, the fracture growth and propagation are similar to the previous case, where one major crack propagates throughout the system and only a few short side branches occur. As α increases, the side branches become larger and larger and many cracks are formed as the external stress increases and the macroscopic failure point approaches. Here the broken bond or spring clusters are no longer linear (as in the previous cases or for cases with $\alpha \cong 0$), and a fractal-like cluster of broken bonds is formed before the global failure. In their two-dimensional simulations with $\alpha = 1/2$ and $3/4$, Sahimi and Goddard (1986) observed that the appropriate fractal dimension d_f of the broken cluster before macroscopic fracture becomes $d_f \cong 1.55 \pm 0.25$. Although a large scatter in this value is observed in different realizations, the structure of the broken clusters is statistically never close to a linear one and the average fractal dimensionality of the cluster is considerably higher than unity.

3.6.2 Models with random breaking strengths

Sahimi and Arbabi (1993) considered the fracture properties of an elastic network where each bond has a fixed spring constant (κ) but the critical or threshold strain or the bond-length a_c is randomly distributed following the distribution

$$P(a_c) \sim (1-\alpha)a_c^{-\alpha}, \quad \alpha \ll 1. \tag{3.25}$$

They considered both central force as well as the bond-bending models. In the latter (bond-bending model), the elastic energy (H) of the network is contributed both by the central force (spring constant) part and the

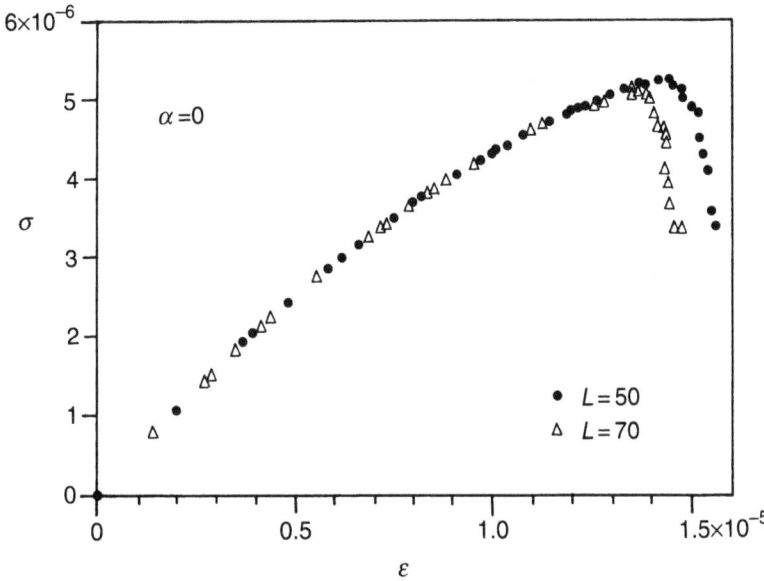

FIG. 3.14. Computer simulation data for stress (σ) - strain (ϵ) in fractured triangular network with bond-bending forces (with $\beta = 0.1$) for two different linear sizes L of the sample. The bond breaking strain distribution is assumed to be uniform ($\alpha = 0$) (Sahimi and Arbabi 1993).

bond-bending part. H is represented here by (1.11) of Section 1.2.1(f). The fracture dynamics of such networks can be simulated in two ways: one in which the bond breaking occurs when the strain l of a bond goes beyond its threshold value a_c, and in the other the bond breaks when its strain energy fa goes beyond its threshold value $\kappa_c a_c^2$, where f is the total microscopic force on the bond.

Since random and continuous bond breakings occur in such models with continuous bond failure thresholds (eqn (3.25)), at the initial level of external stress or strain the crack propagates at a relatively slow rate (see also Herrmann et al. 1989). However, the breakings occur essentially in a fractal-like way, with significant branchings appearing. Very soon the fracture process becomes intense and reaches a maximum. The stress (σ) - strain (ϵ) curve typically looks like that in Fig. 3.14, where the results for the bond-bending (with $\beta = \kappa_b/\kappa_c = 0.1$ in Hamiltonian (1.11)) model on triangular networks of sizes ($L = 50$ and 70) are shown for $\alpha = 0$ in (3.24) (Sahimi and Arbabi 1993). The quantitative features of the stress-strain curves are in perfect agreement with the experimental observations in various kinds of concrete etc. (Brace and Orange 1988). Quantitatively, it is

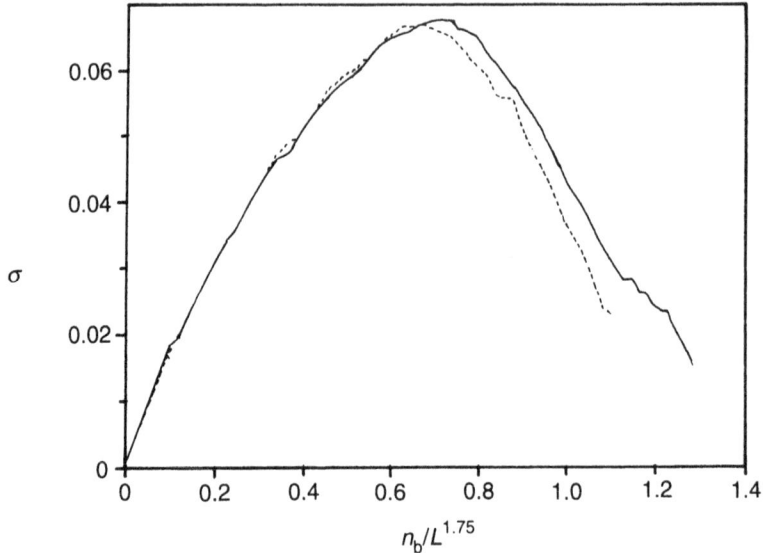

FIG. 3.15. Collapse of the computer simulaion data for stress (σ) - strain ($n_b/L^{\Omega''}$) in fractured triangular network with bond-bending forces ($\beta = 0.1$) for two different linear sizes L (= 50 and 70). The randomness in the bond-breaking strength is assumed to be uniform ($\alpha = 0$). n_b denotes the number of broken bonds and Ω'' is taken to be 1.75 here (Sahimi and Arbabi 1993).

seen that the stress-strain data for various system (lattice) sizes fit very well with the scaling form

$$\sigma \sim \left[\frac{L^\Omega}{(\ln L)^\psi}\right] h\left(\frac{\epsilon}{L^\Omega}\right), \qquad (3.26)$$

where the exponents Ω and ψ are found to be insensitive to the breaking strength distribution (3.25); Ω and ψ are observed to be independent of α. Sahimi and Arbabi (1993) obtained in their simulations

$$\Omega \cong d - 1 \cong 1 \pm 0.1, 2 \pm 0.1$$

and

$$\psi \cong 0.1, 0.2 \qquad (3.26a)$$

for $d = 2$ and 3 respectively. This suggests that the surface area of the sample ($\sim L^{d-1}$) over which the stress is applied determines the scaling of the stress, as well as of strain, since the elastic modulus is scale (L)-independent. Sahimi and Arbabi (1993) also studied the scaling variations of stress σ with the number of broken bonds n_b (related to strain ϵ) in such systems during the fracture process. For the bond-bending model on a triangular network, the scaling collapse of the data for system sizes $L = 50$ and 70 (see Fig. 3.15) gave

$$\sigma \sim L^{\Omega'} g\left(\frac{n_b}{L^{\Omega''}}\right) \quad (3.27)$$

with $\Omega' \cong \Omega \cong 1 \pm 0.1$ and $\Omega'' \cong 1.7 \pm 0.1$. Simulation results of de Arcangelis et al. (1985) and Arcangelis (1990) for two-dimensional solids, formed of a lattice of beams (instead of springs), gave $\Omega' \cong \Omega \cong 0.75$ and $\Omega'' \cong 1.7$ (see also Herrmann 1990, Hansen et al. 1990, 1991). These results indicate the fractal dimension of the resulting fracture or crack to be comparable to that of a random diffusion-limited aggregation process ($n_b \sim L^{d_f}$; $d_f \cong \Omega'' \cong 1.7$ in $d = 2$). Also this value of Ω'' compares well with the results obtained from the data for fracture surfaces of rocks at small scales (Sahimi and Arbabi 1993). Sahimi and Arbabi also observed that the fractal dimension of the sample spanning broken-bond cluster is much smaller ($\cong 1.2$ for $d = 2$, which suggests the backbone dimension d_B to be around 1.2).

For the distribution of fracture strength in such systems, the scaling fit was better obtained for the Weibull distribution (power law with sample size variations) than with the double exponential Gumbel distribution (Sahimi and Arbabi 1993).

3.7 Dynamics of fracture: scaling behaviour for fracture growth patterns and propagation

So far, we have discussed the static aspects (or at best, the quasi-static aspects) of fracture strength, indicated essentially by the initiation or nucleation of fracture (additional breakings over the initial crack), due to stress on such disordered solids. As discussed earlier in Section 3.3, the fracture of disordered solids, after its initiation, is seen to have some interesting growth properties in shape and forms (Mandelbrot et al. 1984, Bouchaud et al. 1993a,b, Mosolov 1993, Roux 1994, Kertesz 1992, Meakin 1990, Maloy et al. 1992, Brechet et al. 1993, Larralde and Ball 1995, Rautiainen et al. 1995, Ray and Date 1996). The fracture propagation has some very interesting and intriguing (often random or chaotic) dynamics of its own.

3.7.1 Fracture propagation velocity

In fact, the first quantitative attempt to incorporate the dynamics of the crack into the Griffith energy balance concept was given by Mott (1948). He suggested that unlike the Griffith case of fracture initiation or nucleation in brittle materials, the kinetic energy of the fracture propagation will have to be extracted out of the released elastic energy. As discussed in Section 3.2.2, the total elastic energy released from a brittle solid with a linear crack of size l is $E_l \sim (\sigma^2/2Y)l^d$, out of which the surface energy required for the creation of the crack surfaces is $E_s \sim \Gamma l^{(d-1)}$. Unlike the Griffith case for fracture initiation, where E_l was equated with E_s, the dynamics of fracture propagation, if any, will require the additional kinetic energy term E_k to be extracted out of the same E_l: $E_l = E_s + E_k$.

Typically, the kinetic energy E_k of fracture propagation can be estimated as

$$E_k \sim \frac{1}{2}\rho \int \dot{u}^2 dV$$

$$\sim \frac{1}{2}\rho \dot{l}^2 \int \left(\frac{du}{dl}\right)^2 dV, \qquad (3.28)$$

where u denotes the displacement of the crack surface, ρ denotes the mass density of the solid and the volume element is denoted by dV. For a brittle solid, the displacement u of a crystal surface will be of the order of $\sigma l/Y$ and the above integral is of the order of $l^2(\sigma/Y)^2$ on dimensional grounds. Hence, the energy balance gives the equation (Mott 1948)

$$a\frac{\sigma^2 l^2}{Y} \sim b\Gamma l + \rho l^2 \dot{l}^2 \left(\frac{\sigma}{Y}\right)^2$$

or

$$v = \dot{l} \sim \left(a\frac{Y}{\rho} - b\frac{\Gamma Y^2}{l\sigma^2}\right)^{1/2}, \qquad (3.29)$$

for the velocity of the fracture propagation (where a and b are some numerical factors depending on the geometry of the crack). This shows that as the crack length increases, the fracture propagation velocity approaches the velocity of transverse sound in the solid, apart from some numerical factor. Thus, if the material is not ductile, the fracture propagation velocity eventually approaches the velocity of sound in the material, and is not dependent on the applied stress or the surface energy density of the material, etc.

The above results are all for a perfect solid under stress, with a single microcrack inside. For randomly disordered solids, the appropriate modification of the above Mott formula has not been developed yet. However, some qualitative features of the fracture propagation process in extremely disordered solids, like the percolating solid near its percolation threshold, are quite obvious and interesting. Although the (equilibrium) strength σ_f of the solid vanishes near the percolation threshold p_c ($\sigma_f \sim (\Delta p)^{T_f}$), the

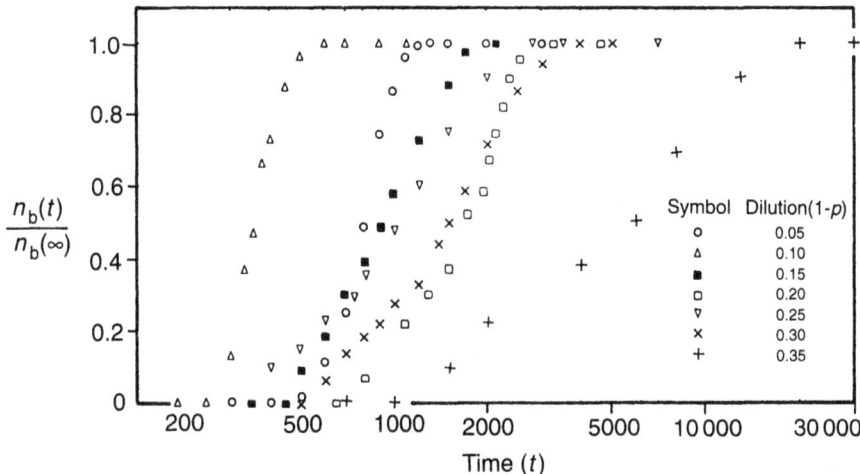

FIG. 3.16. Variation with time t (in number of iterations) of the molecular dynamic simulation data for the number n_b of broken bonds in a given configuration of site dilute Lennard-Jones system (of linear size $L = 21$), for different initial concentraions of dilution, when subjected to a stress just greater than the corresponding fracture stress σ_f (Chakrabarti et al. 1986).

fracture propagation velocity also vanishes near p_c ($v \sim \sqrt{Y/\rho} \sim (\Delta p)^{T_e/2}$ from (3.29)). This indicates the possibility of the intriguing property of a dynamic strength of such solids for short time scales (of the order of $\xi/v \sim (\Delta p)^{-(\nu + T_e/2)}$), the final (equilibrium) strength being vanishingly small.

Chakrabarti et al. (1986) studied the growth of fracture in a randomly (site) diluted Lennard-Jones system on triangular lattice (see Section 3.4.1 (a)), by measuring the total number of broken bonds n_b as a function of time t, when the applied force is slightly greater than the failure strength of the sample. However, the velocity of any single crack (if defined) could not be obtained from such a study. The overall growth of n_b with time was in fact found to be exponential; a clear indication of slowing down (perhaps critical) was observed as the initial dilution concentration approached the percolation threshold: $n_b \sim \exp(t/\tau)$, where τ grows as p approaches p_c (see Fig. 3.16).

3.7.2 *Large propagation velocity and morphology of fractured surfaces*

As mentioned earlier, although the fracture propagation velocity can theoretically approach the transverse sound velocity in the solid, in practice

FIG. 3.17. Molecular dynamic simulation results for the onset of fracture growth instablity in a triangular lattice network with Lennard-Jones potential, having an initial crack at the left-side boundary. (a) Initial stages of growth, and (b) late stage unstable growth with large propagation velocities (Abraham et al. 1994).

it remains much lower than that. Because of the nonlinearities involved in the dynamical equations for fracture propagation (Langer 1993, Langer and Nakanishi 1993, Marder and Liu 1993), at high propagation velocities, various instabilities arise at the crack-tip when the tips split and oscillate (eventually randomly even for perfect solids) and the crack propagation creates random surfaces. However, several recent observations indicate that although the dynamics of failure is intrinsically nonlinear (and also dissipative due to irreversibility) the dynamics of growth is non-chaotic. This has also been confirmed in a recent experiment (Yuse and Sano 1993) on the dynamics of the cracks in thin glass plates with thermal stresses, where the dynamics seems to undergo a sequence of numerous but reproducible instabilities, that are not sensitive to every detail of the fluctuations in the initial conditions.

Recently, various molecular dynamic simulations of triangular lattice Lennard-Jones systems (as discussed in Section 3.2.3) have also been able to reproduce these instabilities, inducing random fracture growth even for non-random systems (Abraham et al. 1994). Figure 3.17 shows such a growth of fracture for a Lennard-Jones system on triangular lattice (using about 10^6 atoms). These two-dimensional simulations showed that the maximum attainable crack-tip speed is considerably smaller (about 32% of the sound velocity in the model), and the roughness of the fractured surface (at this maximum speed of propagation) is considerable. Abraham et al. found a very high value of the roughness exponent: $\zeta \simeq 0.8$ in two dimensions. Such roughness in the morphology of the fracture surface due to branching instabilities at higher velocities of fracture propagation can also be seen from the computer simulation of pure lattices (Marder and Fineberg 1996), as shown in Fig. 3.18. This indicates that pulling harder on a crack can often slow it down by spreading the fracture in the transverse directions. In fact, similar simulations for amorphous structures, by Nakano et al. (1995), indicated the interesting feature that the roughness of the cracked surfaces, produced during the fracture propagation, crosses over from a relatively small value for the roughness exponent ($\zeta \simeq 0.4$) to a comparatively larger value ($\zeta \simeq 0.8$; see Section 3.3) as the crack velocity increases with its growth in size (see Fig. 3.19).

3.7.3 *Elastic precursor effects of complete fracture*

Another interesting feature of the fracture propagation, indicating a universal aspect, was observed by Sahimi and Arbabi (1992). In their computer simulation study of fracture in triangular lattice networks with both central and bond-bending forces (having force constants κ_c and κ_b respectively in Hamiltonian (1.11)), with the distribution (3.25) for the threshold strain a_c of any bond, they measured some linear response (elastic moduli) of the solid, as the fracture propagates. Specifically, they measured the ratio $r = C_{11}/\mu$ of the compressional to the shear modulus, as the fracture propagates under the action of the external stress, on different classes of solids (different $\beta = \kappa_b/\kappa_c$). In fact, this ratio was already suggested (Bergman and Kantor 1984, Bergman 1985), and later shown, to approach a universal value for percolating networks as p goes to p_c. It was therefore natural to investigate the behaviour of this quantity in the case of fracture propagation. With uniform distribution of the threshold strain ($\alpha = 0$ in (3.25)), Sahimi and Arbabi observed that as more and more bonds break (for their respective strain going beyond their threshold value a_c), the ratio r appears

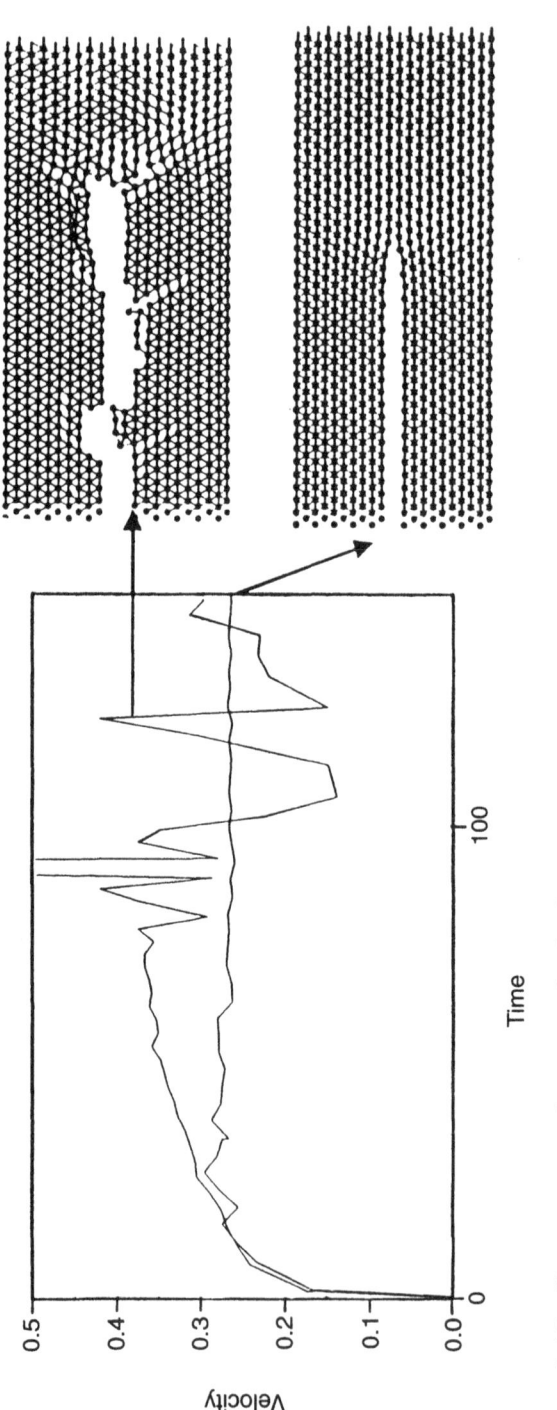

FIG. 3.18. Computer simulation results for fracture growth in perfect lattice. Transition from smoothly advancing crack to violent propagation and branching instabilities occurs with larger pulling stresses (from Marder and Fineberg 1996).

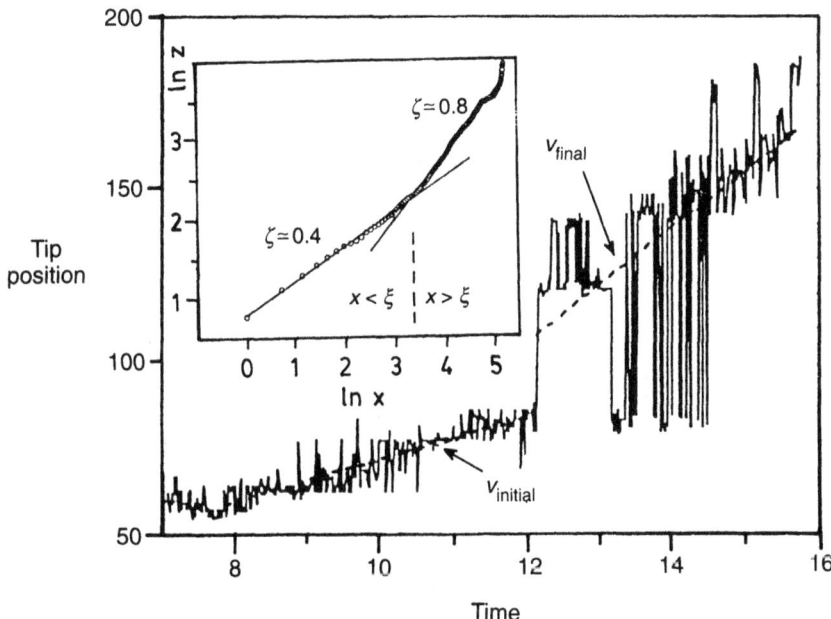

FIG. 3.19. Molecular dynamic simulation results for the fracture propagation in amorphous structures (with Lennard-Jones potential) show that the average fracture velocity crosses over to a higher value (v_{final} from v_{initial}, indicated by the dotted lines) at the late stages of growth, as the crack size exceeds the typical size (correlation length) ξ of the voids in the network. The inset shows that a corresponding crossover in the fractured surface roughness exponent also occurs along with the crossover in the fracture velocity (from Nakano et al. 1995).

to approach a universal value ($\simeq 1.25$ for such two-dimensional isotropic elastic networks), irrespective of their initial values for vanishing external stress. This is shown in Fig. 3.20. As can be seen from the figure, this feature is true even for the central force system ($\beta = 0$). It is seen that for small amount of local fracture or large fraction of unbroken springs, the ratio r remains practically constant and then starts changing drastically, as the macroscopic breakdown point approaches, when the broken bonds almost percolate.

Such behaviour of fracture propagation can clearly be of extreme importance for practical purposes. For example, the fraction r, from the measurement of the sound velocity ratio of compressional and shear waves, can indicate the proximity of the imminent macroscopic failure or fracture of

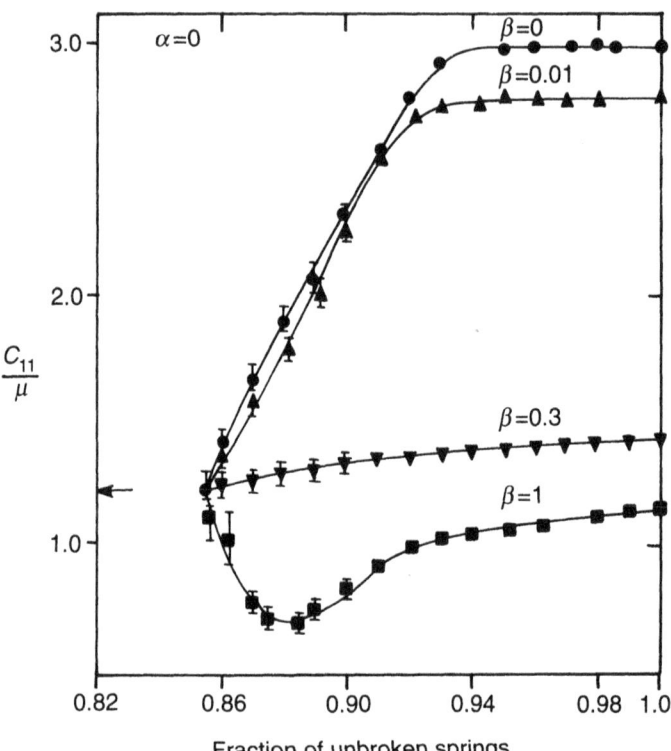

FIG. 3.20. Computer simulation results for the ratio C_{11}/μ versus the fraction of unbroken springs in a triangular lattice network with different bond-bending forces ($\beta = 0$, 0.01, 0.3 and 1), having uniform distribution of bond-breaking thresholds. The ratio C_{11}/μ seems to converge to an universal value ($\simeq 1.25$) as the complete fracture point is approached (Sahimi and Arbabi 1992).

the entire sample.

Recently, Sahimi and Arbabi (1996) have observed that the cumulative elastic energy E_r released during fracturing of heterogeneous solids follows a power law variation with the strain ϵ of the solid, having a log-periodic correction:

$$E_r(\epsilon) = A + B\epsilon^m[1 + C\cos(D\ln\epsilon + E)], \qquad (3.30)$$

where A, B, C, D and E are constants, and m is an exponent. In a computer simulation study, similar to that above, for an elastic network of size 80×80, with central forces described by the Hamiltonian (1.11) with $\beta = 0$ and the threshold strain distribution for each bond given by (3.25)

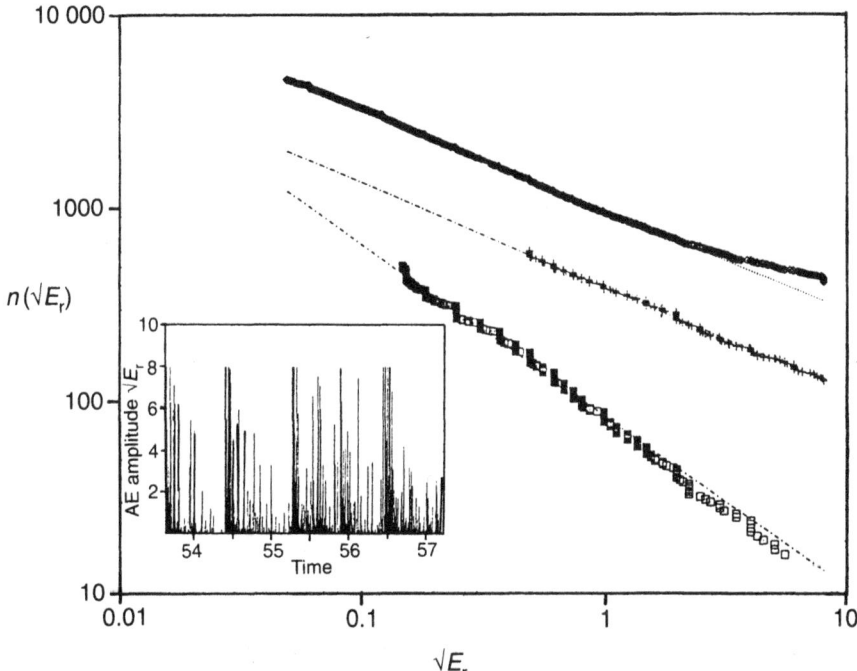

FIG. 3.21. Cumulative distribution of acoustic emission bursts with respect to their amplitudes for three different samples. The inset shows a typical experimental record of the acoustic emission signal time series, with the propagation of fracture. The five stronger bursts correspond to increases in the stress (from Petri et al. 1994).

with $\alpha = 0.5$, they observed the above relationship. In particular, they observed that the released elastic energy E_r (normalised by the maximum elastic energy E_0 before any fracture) due to the propagation of the fracture varies with the strain ϵ (normalised by the strain ϵ_0 just before the sample spanning fracture) following the above law (3.30), with $m \simeq 0.13$.

This observation is consistent with the scaling of acoustic emissions preceding the fracture of composite materials with the time (replaced here by the strain), and may thus provide a basis for predicting the fracture of materials. It may be noted that such power laws (with again the log-periodic correction) were in fact observed earlier for the scaling of seismic activity with time, before a large earthquake (Bufe and Varnes 1993, Sornette and Sammis 1995).

3.8 Dynamic annealed impurity and self-organised criticality in fracture

As the fracture propagates, the elastic energy released due to the microfractures occurring within the sample can be measured. This ultrasonic emission due to micro-fracture aftershock relaxation has recently been measured for various laboratory samples. Petri et al. (1994) measured the ultrasonic emission amplitude distribution in a large number of stressed solid samples under different experimental conditions. A power law decay for the cumulative energy release distribution $n(E_r)$ with the released energy amplitude E_r was observed in all cases: $n(E_r) \sim E_r^{-c}$ (see Fig. 3.21). This is indeed very similar to the Guttenberg-Richter law for the frequency distribution of earthquakes, as discussed briefly in Chapter 1, and will be discussed in detail in the next chapter.

Several attempts have been made to model such behaviour of fracture growth in disordered solids. Very recently, Caldarelli et al. (1996) have shown that such self-organised critical behaviour (see Section 1.2.3) for the energy releases during fracture growth in disordered solids can be obtained only in presence of some kind of dynamic 'annealed' disorder. In the kinds of disorder that we have considered so far, the disorders are 'quenched', in the sense that positions of the various disordered sites or bonds are fixed and can not change with time (or with the dynamics). Caldarelli et al. showed that the above kind of self-organised critical behaviour, giving the power law for the distribution of the failure size (or the released elastic energy) due to micro-fractures, is possible in a randomly quenched disordered solid, when the breaking strength distribution around the crack can change dynamically (in an annealed way) with the stress concentration.

Caldarelli et al. considered a bond-bending system with Hamiltonian (1.11), in two dimensions, having random quenched breaking thresholds given by the distribution (3.25) with $\alpha = 0$. Local stresses are calculated, and if the local strains (of the springs) exceed the the threshold value of the spring, the spring breaks and the energy of the broken spring gets released. The growth in the number of broken springs with time was also counted. With just this quenched disorder alone, they did not see any self-organised criticality. When they redistributed the breaking strength of the springs, neighbouring the already broken springs, in such a way that the redistributed thresholds are lower than the earlier (stress concentration around the crack weakens the neighbouring springs), they observed intriguing power law behaviour in the failure characteristics. This kind of correlation or annealed disorder is shown to lead to a self-organised critical behaviour in the broken cluster size distribution. Although no such behaviour was seen straightforwardly for the distribution of the elastic energy release, a renormalisation of the released elastic energy (of the broken bonds) by the total stored potential or elastic energy (of the unbroken springs) showed similar power law behaviour (Caldarelli et al. 1996).

3.9 Summary and conclusions

We have modelled the random elastic solids as percolating networks of springs, having concentration p (the rest of the springs being cut or absent). When the energy contributed by the springs comes only from the changes in their respective length, the network is a central force system. Such regular spring networks are unstable on hypercubic (e.g. square, simple cubic, etc.) lattices. Additional contribution of the bond-bending force can stabilise such a network (apart from randomness in spring lengths). With random dilution (randomly cut springs), the elastic properties of such stable networks are well studied and understood.

We have studied the the fracture properties of such elastic networks, under large stresses, with initial random voids or cracks of different shapes and sizes given by the percolation statistics. In particular, we have studied the cumulative failure distribution $F(\sigma)$ of such a solid and found that it is given by the Gumbel or the Weibull form (3.18), similar to the electrical breakdown cases discussed in the previous chapter. Extensive numerical and experimental studies, as discussed in Section 3.4.2, support the theoretical expectations. Again, similar to the case of electrical breakdown, the nature of the competition between the percolation and extreme statistics (competition between the Lifshitz length scale and the percolation correlation length) is not very clear yet near the percolation threshold of disorder.

The scaling behaviour of the most probable fracture strength σ_f, expressed by the fracture exponent T_f near the percolation threshold, has been investigated extensively. The theoretical results compare well with those observed experimentally, and in computer simulations. Although considerable progress has been made here, experimental results for continuum percolation are scarce, and more investigations are clearly necessary.

The problem of the dynamics or growth of fracture in such randomly disordered solids, after the initiation or nucleation of fracture, is very intriguing. In particular, the knowledge about the nature of the self-organised behaviour for fracture propagation in disordered systems (as discussed in Section 3.7) and its criticality (if any), is potentially very important. The predictability of macroscopic fracture of such solids requires extensive understanding about the dynamics of fracture. The results of the investigations, so far (discussed briefly in Sections 3.6 and 3.7), are very rudimentary. However, the subject is making tremendous progress recently and considerable developments and understanding are expected soon.

4
EARTHQUAKES IN MODEL SYSTEMS

4.1 Introduction

We discuss in this chapter several model studies for earthquakes. Earthquakes are large scale failures, occurring mainly as a consequence of frictional slips in the upper region of the earth's crust, to adjust the stress accumulated by the large scale tectonic plate motions over long periods of several decades of years (see e.g. Guttenberg and Richter 1954, Kostrov and Das 1988, Scholz 1990). Like in the cases of fracture or breakdown (discussed in the previous chapters), where the stresses or fields are redistributed within the surviving or unbroken portions of the material, the stresses on the sticking tectonic plates are readjusted after each individual slip in any region (plate). However, unlike in the fracture or breakdown case where a fractured, broken or fused bond is permanently broken and remains unable to sustain any stress any further, the slipped or failed blocks in the earthquake can again stick and participate again in sustaining stresses. An earthquake occurs when a large number of the blocks of the earth's crust slip simultaneously, releasing a large amount of elastic energy. Such stick-slip frictional instability of a large number of blocks, as in the case of earthquakes, is a typical dynamic failure phenomenon.

Earthquakes are in fact surface manifestations of the phase changes and the consequent deformations in earth's mantle, occurring at depths of hundreds of kilometres where the temperature and pressure are around a thousand degrees celsius and gigapascals respectively. The earth's solid outer crust rests on the tectonic shell, having typical thickness of the order of some tens of kilometres. This tectonic shell is divided into a small number of mobile plates or blocks (about twelve in number; see Vilotte et al. 1994), having relative velocities of the order of a few centimetres per year. These dynamic instabilities, resulting from the relative motions of the tectonic plates, are settled through various slip events, releasing the stored elastic energy. During an earthquake, the slip velocity (relative between the two edges of the failure or fault plane) becomes of the order of few metres per second. Numerous observations suggest a power law distribution for the elastic energy released during earthquakes. This empirically observed power law distribution of event sizes is known as the Guttenberg-Richter law (Guttenberg and Richter 1954), which is considered to be one of the

Introduction

most fundamental observations in geology and seismology. This law states that the number n of earthquakes of magnitude greater than or equal to m (on the Richter scale) is given by

$$\log n = \text{Constant} - am, \qquad (4.1a)$$

where the value of the constant depends on the location, the value of a being generally within the interval $0.8 < a < 1.5$. The (elastic) energy (E_r) released during an earthquake is assumed to increase exponentially with the earthquake magnitude m:

$$\log E_r = \text{Constant} + bm, \qquad (4.1b)$$

where b varies from 1 to 1.5. In fact, the above equation (4.1b) defines the earthquake magnitude. Combining the above equations, one gets the Guttenberg-Richter power law for the frequency distribution of energy:

$$\log n = \text{Constant} - c \log E_r,$$

or,

$$n \sim E_r^{-c}. \qquad (4.1c)$$

The recorded data indicate that $c = b/a$ varies from 0.8 to 1.1. It may be mentioned that Guttenberg and Richter obtained this result from averages over the earthquake events observed throughout most of the world, and not for just one earthquake fault.

In the last 10-15 years, several hypotheses and model systems have been studied to investigate the nature of the earthquake phenomena. Most of these model studies are intended to capture the Guttenberg-Richter type power law for the frequency distribution of the failures. Among these, the majority of the models are of the stick-slip kind. Here the solid-solid friction occurs due to the shear force of the 'junctions', where the two solids are in real contact. This contact area effectively decreases with the increasing relative velocity during slip, which gives rise to velocity-weakening frictional forces (see e.g. Bowden and Tabor 1950, Heslot et al. 1994). In any region, the earth's crust touches and rests on one of the moving tectonic plates at various points or 'junctions', which provide the sticking frictional force to follow the motion of the mobile plate. The elasticity of the earth's crust provides the restoring force. This competition between the sticking frictional force and the restoring elastic force necessitates stabilisation of the dynamics of the system through various slips, which releases the stored additional elastic energy. The models in this class are intended to capture such dynamics. These are all prototypes of the original Burridge-Knopoff model (Burridge and Knopoff 1967). Attention is focused here on the dynamics of discrete fault models, consisting of massive and rigid blocks connected by springs supported on a horizontal rough solid platform, which moves constantly with a small velocity. With the velocity-weakening friction force (between the blocks and the platform), the study of the complex dynamics of the blocks gives many new perspectives, and in some cases a Guttenberg-Richter type power law for the slip events. These studies have recently been revitalised by Carlson and Langer (Carlson and Langer 1989a,b, Carlson et al. 1994), who studied numerically some simplified versions of this model. Their model studies indicate that in some, not very unique, regime (depending on the parameter values of the spring constants and the velocity-weakening frictional force) such models give rise to complex failure sequences suggestive of the Guttenberg-Richter law.

In an extremely simplistic cellular-automata version of such models, Bak et al. (1988) and Bak and Tang (1989) considered discretised automata values representing the stress on any block (which in this cellular-automata model is represented by a cell). Any automaton value then increases uniformly with time, and also gets its (equal) share from its neighbouring cell which fails or slips when its automaton value (stress level) exceeds a preassigned threshold (see the sandpile automata models discussed in Section 1.2.3). Once such failure occurs in any cell, its automaton value is set at zero, from which it starts again to build up. Both analytic and numerical solutions of this kind of dynamics give a power law distribution for the failure events in a very natural way, coming from the self-organised criticality for the dynamics of slip events. These models indicate the origin of the Guttenberg-Richter law to be due to the self-organised critical behaviour of the earthquakes.

In another recent trend of such investigations, one considers the Guttenberg-Richter (power) law to result as a consequence of the criticality of the geometry of the earthquake (fracture) faults of the earth's upper crust, where established power law distributions for the fault geometries occur. One then compares with those for percolation clusters near or at the percolation threshold. One therefore investigates the fracture mechanics of the stressed earth's crust, where such fractal patterns for the fault segments occur near the contact areas of the major plates (Kagan 1982, Barriere and Turcotte 1991, Sahimi 1992).

4.2 Burridge-Knopoff stick-slip model of earthquakes

4.2.1 Laboratory simulation model

In their original one-dimensional model for the 'table top' laboratory simulation of earthquakes, Burridge and Knopoff (1967) took a chain of eight massive wooden blocks (of mass around 140 grams each) connected by iden-

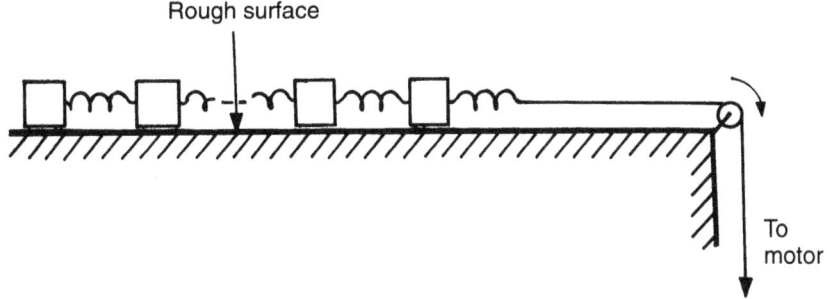

FIG. 4.1. Schematic diagram of the Burridge-Knopoff model of earthquakes.

tical coil springs (with force per unit strain of the order of 2 newtons) with the neighbouring block. The first block was connected to a string through another such spring, and the string was pulled using a driving motor. The blocks and the springs were placed on a horizontal rigid platform with rough surface (like sand-paper), and the driving motor was used to pull the first spring horizontally as shown in Fig. 4.1.

Starting with unstrained springs, when the blocks are pulled slowly by the string at constant rate, using the motor, the blocks initially stick to the surface (show a little bit of creep motion along the platform) until they slip, when the spring strains get readjusted. This elastic spring chain, together with the velocity-weakening friction force here, gives rise to intriguing complex dynamics of the system. Noting the instantaneous positions of the blocks, the elastic energy $E_T(t)$ of the chain of N springs (and blocks) can be calculated using

$$E_T(t) = \frac{1}{2} \sum_{i=1}^{N} \kappa [(x_{i-1}(t) - x_i(t) - l_0)]^2,$$

where κ is the spring constant of the springs connecting any two blocks, l_0 is the unstretched length of the springs, and $x_i(t)$ denotes the linear position coordinate of the ith block at time t. Figure 4.2 shows a typical variation of the energy E_T with time. It clearly shows the charging due to constant strain during the sticking period of one block and the forward slip of the neighbouring block, and the subsequent releases of energy during slips of neighbouring blocks. Initially, the system takes some time, depending on the uniform strain rate, to charge or get strained up to a limit where each spring gets strained (stretched) up to the limit of the frictional stability of the blocks. Although individual small slips occur here and there during

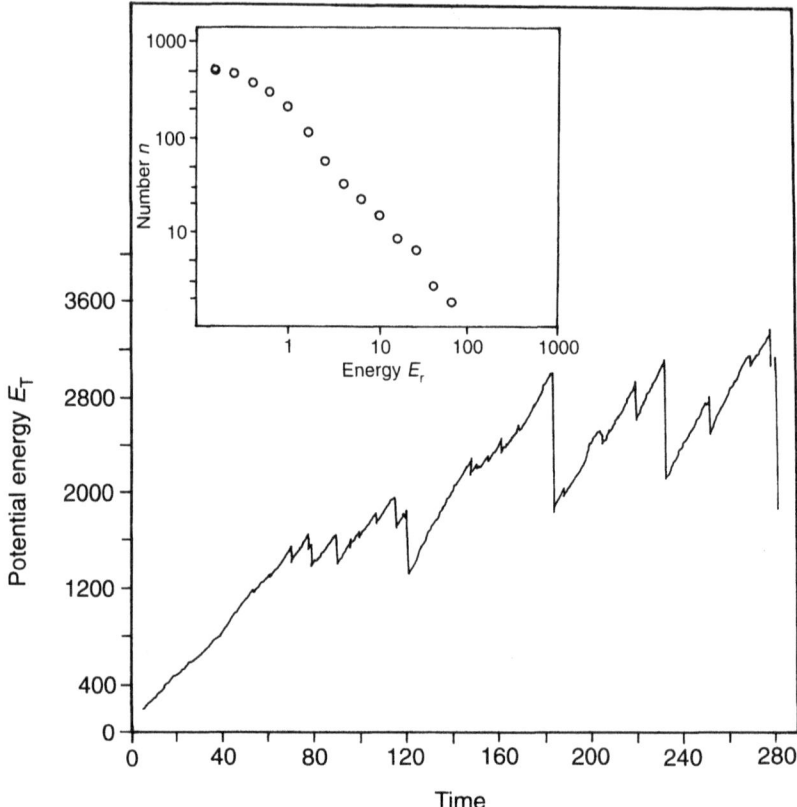

FIG. 4.2. A typical plot of the time variation of the potential energy (E_T) of the spring-block model. Each discontinuity (almost vertical fall) in the potential energy correspond to an 'earthquake' with the magnitude given by the energy release (E_r) corresponding to the fall. A plot of the frequency $n(E_r)$ versus energy release (E_r) on a log-log scale is indicated in the inset (cf. Burridge and Knopoff 1967).

this charging period, eventually after the charging, big events of collective slips of almost all the blocks occur. These big events release a large amount of elastic energy (indicated by the almost vertical drops in energy in Fig. 4.2). After such a big event of collective slips, the system charges again for some time (with individual and isolated slips here and there), and then fails again macroscopically. This quasi-random sequence of large scale slip events is identified in this model as (major) earthquakes. In fact, Burridge and Knopoff showed that the number n of such 'earthquakes' in the model,

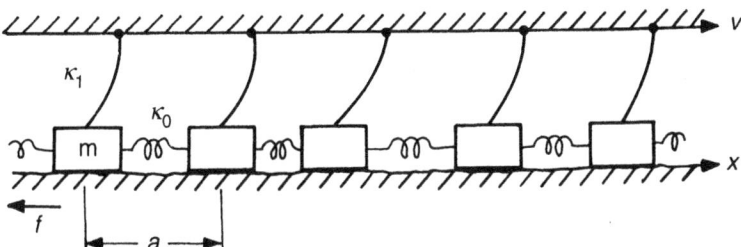

FIG. 4.3. A simple spring-block model in one dimension. The system is spatially homogeneous, composed of equal masses m for each block, each connected to nearest neighbours (separated in equilibrium by a distance a) by springs of equal strength κ_0, and to a uniformly moving surface (with velocity V) with springs of equal strength κ_1. The system rests on a rough platform, which exerts a friction force f in the direction opposite to the relative motion and magnitude depending on the relative velocity of the block.

with released elastic energy greater than or equal to E_r, varies indeed by a power law, as shown in the inset of Fig. 4.2, which suggests $n \sim E_r^{-c}$, with $c \sim 1$.

4.2.2 Computer simulation model

Although Burridge and Knopoff also considered numerical solutions of such systems, the recent upsurge in the interest in numerical studies of the Burridge-Knopoff model started with the work of Carlson and Langer (1989a,b). The minimal model considered here contains a linear array of N blocks, each coupled to its nearest neighbours by elastic springs of identical strength κ_0 and also connected to a rigid support at the top by a different set of elastic springs having identical strength κ_1, as shown in Fig. 4.3. The blocks are placed on a horizontal platform with rough surface, which moves with a uniform velocity V. The equation of motion of the jth block of the system is then

$$m\ddot{x}_j = \kappa_0(x_{j+1} - 2x_j + x_{j-1}) - \kappa_1 x_j - f(\dot{x}_j - V), \qquad (4.2)$$

where the dots denote differentiation with respect to time t, m denotes the mass of any of the blocks, and f represents the nonlinear velocity-weakening friction force. Here, x is measured along the chain length and the pulling velocity V of the platform is also in the increasing x direction (Fig. 4.3). The first term on the right hand side of (4.2) comes from the elastic restoring force of the neighbours, while the next term comes from the elastic restoring force of the rigid support. The third term represents

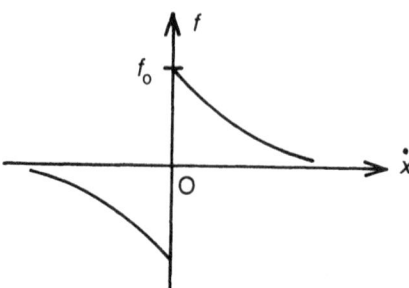

FIG. 4.4. The stick-slip velocity-dependent friction force f as a function of the velocity $\dot{x}$. The force ranges between $\pm f_0$ at zero velocity and decreases monotonically to zero as the velocity magnitude $|\dot{x}|$ becomes large.

the solid-solid friction, which is dependent on the relative velocity of the block with respect to the platform. This friction force is assumed to have the form

$$f(\dot{x}) = f_0 \phi\left(\frac{\dot{x}}{V_0}\right), \quad \text{with} \quad \phi(y) = \frac{\text{sgn } y}{1 + |y|}, \quad (4.3)$$

where V_0 denotes some characteristic speed for the scaling of the friction force. The typical form of $f(x)$ is shown in Fig. 4.4. Thus the friction force ranges between $\pm f_0$, which occurs at zero velocity, and decreases monotonically as the velocity increases ($f(\dot{x})$ becomes $\pm \frac{1}{2} f_0$ at the characteristic speed V_0). This nonlinear nature of ϕ causes the system to undergo discontinuous stick-slip events.

Using appropriate units, the above dynamical equation can be written in terms of dimensionless variables. Although there is no intrinsic length scale in the problem, one can use the unit for measuring the displacements x as $x_0 = f_0/\kappa_1$, which is the maximum displacement of any block. Using this, we define the dimensionless displacements $U_j = x_j/x_0 = (\kappa_1/f_0)x_j$. The time variables can also be scaled using $\tau = t/T$, where $T = \sqrt{m/\kappa_1}$ gives the period of oscillation of a single block in absence of friction. The scaled form of the equation (4.2) is then

$$\ddot{U}_j = \kappa(U_{j+1} - 2U_j + U_{j-1}) - U_j - \phi(\alpha[\dot{U}_j - v]), \quad (4.4)$$

where $\kappa = \kappa_0/\kappa_1, v = V/V_0$ and $\alpha = x_0/(V_0 T)$. Here, the dots denote differentiation with respect to τ.

One can now study approximately, following Carlson and Langer (1989b), the instabilities of some trivial solutions of eqn (4.4). To do so, one can take the continuum limit of the equation, taking the inter-block equilibrium unstrained separation $a \to 0$: $s = ja$ and $\gamma = a\sqrt{\kappa}$, giving

$$\ddot{U} = \gamma^2 \frac{\partial^2 U}{\partial s^2} - U - \phi(\alpha[\dot{U} - v]). \tag{4.5}$$

One trivial solution may be taken as $U = -\phi(-\alpha v) = $ constant independent of space and time coordinate. Here, all the blocks are moving uniformly at the pulling speed relative to the rough surface. However, such a solution can be shown to be unstable against small fluctuations. Considering small perturbations of wavelength q, one can study the stability of the solution

$$U = -\phi(-\alpha v) + Ae^{\Omega \tau} e^{iqs}. \tag{4.6}$$

By straightforward substitution of this solution in (4.5), and using the Taylor expansion $\phi(v + \epsilon) \sim \phi(v) + \epsilon \phi'(v)$, one gets

$$\Omega = \frac{1}{2}[\tilde{\alpha} \pm (\tilde{\alpha}^2 - 4 - 4\gamma^2 q^2)^{1/2}],$$

for small q, where $\tilde{\alpha} = -\alpha \phi'(-\alpha v) \sim \alpha$. The existence of the positive real part of Ω shows that any small fluctuation amplifies unboundedly with time τ and eventually such uniform solutions become unstable. Similar instabilities can also be studied numerically for solutions, spatially uniform but having time variations, or for solutions with spatial fluctuations but uniform in time. Here also, one can show that small fluctuations grow arbitrarily (Carlson and Langer 1989b).

It may be mentioned here that a recent study (Vasconcelos 1996) of a simple noncooperative (one-block) model of stick-slip motion (described by eqn (4.2) with $\kappa_0 = 0$ or eqn (4.4) with $\kappa = 0$) shows discontinuous velocity-dependent transition in the block displacement, for generic velocity-dependent friction forces. Naïve generalisation of this observation for the coupled Burridge-Knopoff model would indicate a possible absence of criticality in the model.

4.2.3 Numerical studies and results

One can solve numerically the time differential equation (4.4) employing, for example, the Runge-Kutta method. Using a very small randomness in the initial conditions for the positions of the blocks (fluctuations given by a white noise amplitude of the order of 10^{-3}), and free boundary conditions, the differentials can be estimated using the fourth-order Runge-Kutta method. Such numerical studies indicate various kinds of solutions ranging from earthquake-like solutions (which are almost on the verge of chaos) to pulsed travelling wave or soliton-like solutions (Carlson and Langer 1989a,b,

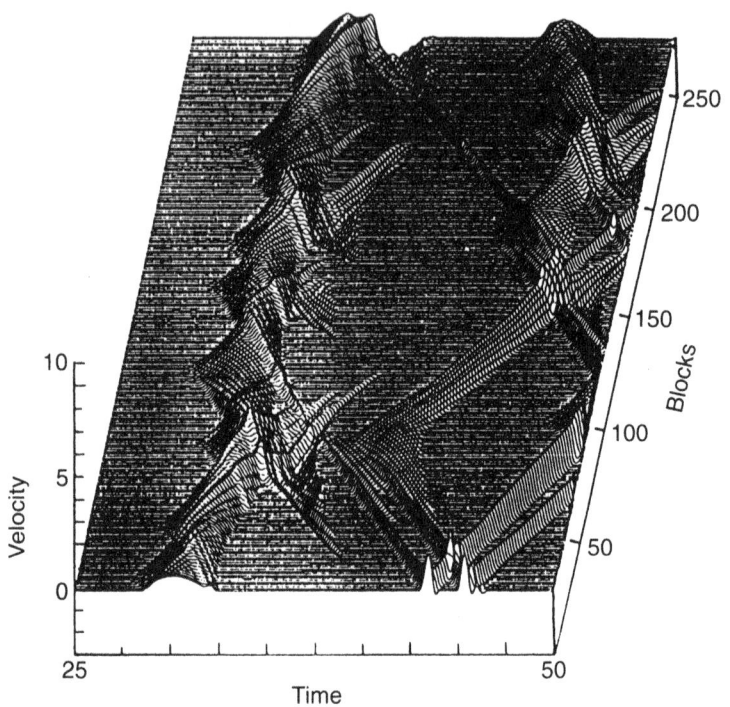

FIG. 4.5. Numerical solution for block velocity against time for each block for a choice of the parameter set values ($N = 250$, $\kappa = 100$, $v = 0.05$, $\alpha = 4$). Large earthquake-like events, involving almost simultaneous motions of all the blocks, can be clearly seen (Ananthakrishnan and Ramachandran 1994).

Carlson *et al.* 1994, Schmittbhul *et al.* 1993, Vilotte *et al.* 1994). In fact, Schmitbhul *et al.* (1993) noted that the system shows earthquake-like almost chaotic solutions for very small values of $\theta = Nv/\sqrt{\kappa}$, while the solutions are soliton-like for θ of the order of unity. This size-dependent transition has been extensively studied by Vilotte et al (1994) and Ananthakrishna and Ramachandran (1994), who also showed that the same transition occurs as the parameter α in (4.4) decreases from higher values. As the blocks in the model represent the independent 'junctions' where the earth's crust rests on a moving tectonic plate (for any particular epicentre region, in case of earthquakes), the number N of such blocks in the model is usually quite small. Also, the (dimensionless) tectonic plate velocity v is very small in reality and of the order of 10^{-8} (Carlson *et al.* 1994). This

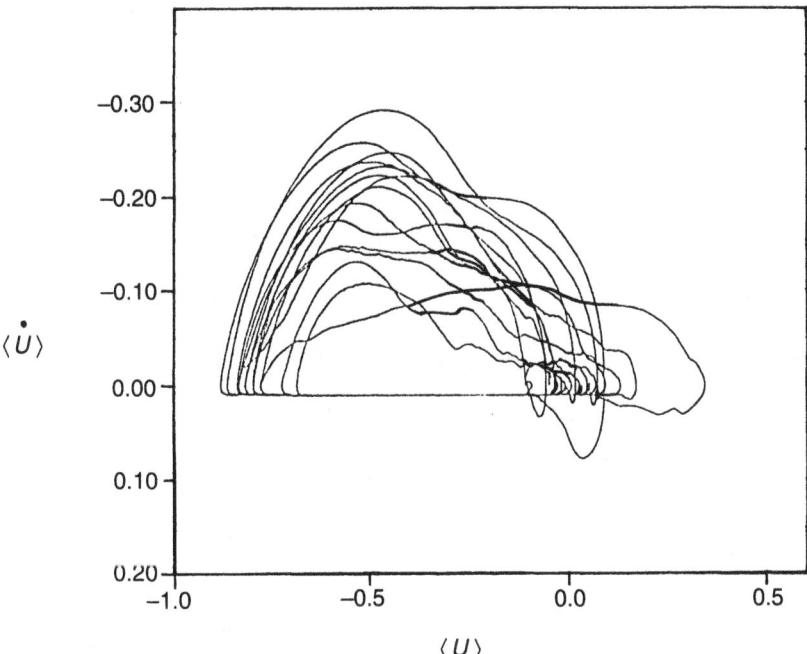

FIG. 4.6. A typical form of the trajectory (limit cycles) showing average (over blocks) velocity $<\dot{U}>$ against the average displacement $<U>$ in the same range of parameter values, where earthquake-like events are observed (Vilotte et al. 1994).

would suggest the appropriate θ value to be quite small, indicating the prominence of earthquake-like solutions for the system.

Indeed, numerical solutions for (4.4) show overall erratic stick-slip motion of the blocks as shown in Fig. 4.5 for $N = 250, \kappa = 100, v = 0.05$ and $\alpha = 4$ (following Ananthakrishna and Ramachandran 1994). Here, events are identified as slipping blocks (blocks with nonzero velocity), and Fig. 4.5 shows several small events involving slip motions of a small number of blocks and one large event involving nearly simultaneous slips of almost all the blocks. In fact, the spatially averaged velocity (averaged over all the blocks) $\langle \dot{U} \rangle$, when plotted against the average displacement $\langle U \rangle$ of the blocks, results in a somewhat random limit cycle. A typical form of the cycle for such earthquakes is shown in Fig. 4.6. This kind of earthquake-like solutions may be compared and contrasted with the similar plot of the velocities of the individual blocks with time, shown in Fig. 4.7, obtained for $N = 250, \kappa = 100, v = 0.05$ and $\alpha = 2.5$. One can clearly see the absence

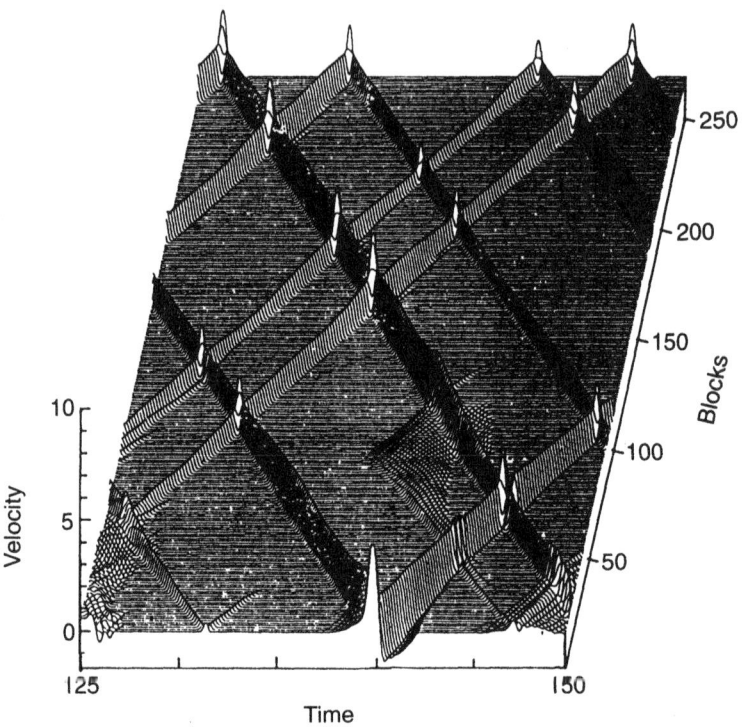

FIG. 4.7. Numerical results for the block velocities against time for each block for a different choice of the parameter set values ($N = 250$, $\kappa = 100$, $v = 0.05$ $\alpha = 2.5$), where earthquake-like events are absent and travelling wave solutions dominate (Ananthkrishnan and Ramachandan 1994).

of large events (earthquakes) and prominence of travelling wave type solutions in this case. In fact, here the plot of average block velocity $\langle \dot{U} \rangle$ versus average displacement $\langle U \rangle$ tends to reduce to a (fuzzy) fixed point (Vilotte et al. 1994). In the small θ and large α limit, the large earthquake events (with nonzero slip velocities for almost all the N blocks) appear to have some well defined statistical recurrence time. In fact, the frequency distribution of large earthquakes (of magnitude over a suitably chosen limit) has a prominent most probable value at a most probable recurrence time dependent on the dynamic parameters (Vilotte et al. 1994).

There are continuous distributions of smaller events, and the cumulative number n of events has a power law variation with the size S of the event, measured by the number of blocks participating in the event : $n(S) \sim S^{-c}$,

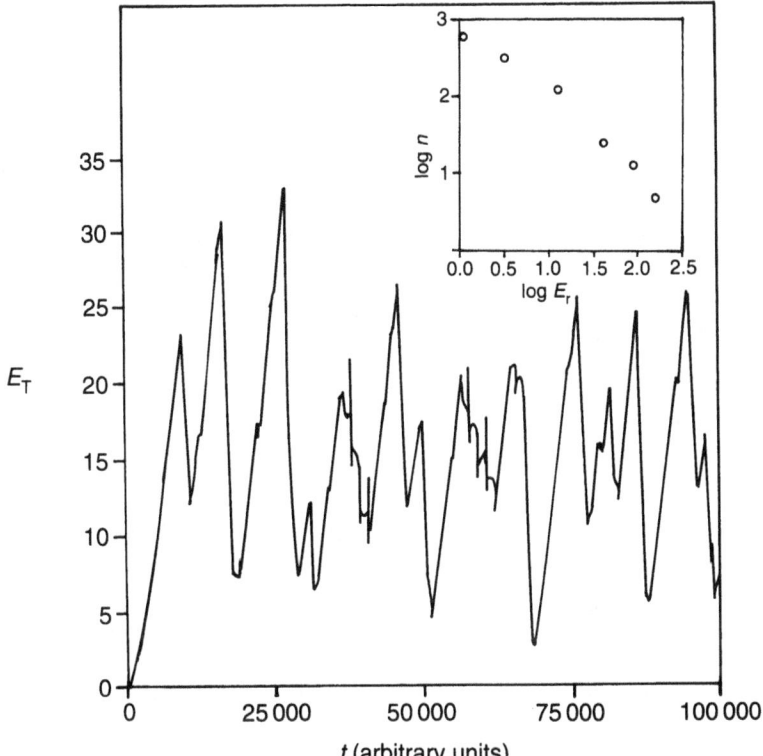

FIG. 4.8. A time series plot of the numerically estimated total potential energy E_T of the spring-block system for $N = 100$, $\kappa = 100$, $v = 0.01, \alpha = 2.5$. The almost vertical fall in E_T gives the amount of earthquake energy release E_r. The inset shows the frequency plot of such energy releases on a log-log scale (Acharyya and Chakrabarti 1995).

with $c \sim 1.1$ (Vilotte et al. 1994). Similar results were also obtained earlier by Carlson and Langer (1989a,b), who measured the magnitude m of the earthquakes by taking the log of the total displacement (strain) of all the blocks and found $n(m) \sim \exp(-am)$, with $a \simeq 1$, in consistence with eqn (4.1a). A similar power law can also be extracted from the time variations of the total elastic energy

$$E_T = (\kappa/2) \sum_j [(U_j - U_{j+1})^2 + (U_j - U_{j-1})^2] + (1/2) \sum_j U_j^2,$$

as shown in Fig. 4.8 for $N = 100, \kappa = 100, v = 0.01$ and $\alpha = 2.5$. The almost vertical fall in energy E_T gives the amount of earthquake energy release E_r, and one can plot the frequency distribution $n(E_r)$ of such energy releases against E_r. The inset of Fig. 4.8 shows the corresponding plot of $\log n(E_r)$ versus $\log E_r$, giving $n(E_r) \sim E_r^{-c}$, with $c \sim 1.2$ (Acharyya and Chakrabarti 1995).

Recently, Myers *et al.* (1996) have studied numerically the two-dimensional version of this model. Again here they observed, for a finite range of the parameter values, chaotic dynamics exhibiting a broad range of earthquake-like events: the frequency-magnitude distribution includes a Guttenberg-Richter type scaling region for the small events, with a slight excess in the frequency for the large events.

4.3 Self-organised criticality and cellular automata models of earthquakes

As mentioned earlier, in Section 1.2.2, the observation of the Guttenberg-Richter law has been suggested to be the manifestation of the self-organised critical (SOC) state of the dynamics of the earthquake faults. Accordingly, several simple cellular automata models have been proposed and studied in this context. One such version of the Burridge and Knopoff model in two dimensions, due to Bak and Tang (1989), is the simple Bak *et al.* (1987) model of a 'sandpile' discussed in Chapter 1 (Section 1.2.3), interpreted appropriately for this purpose. Here, the discrete automaton value $F_{i,j}$ at any cell site (i,j) represents the stress on the corresponding block, and it is assumed to increase continuously with time at random locations. The stress $F_{i,j}$ drops to zero if $F_{i,j} \geq F_0$, and the neighbouring cells (blocks) share the stress:

$$F_{i\pm 1,j} = F_{i\pm 1,j} + 1, \quad F_{i,j\pm 1} = F_{i,j\pm 1} + 1.$$

Here, $F_0 = 4$, the number of nearest neighbours, ensures the stress redistribution with local conservation. The stresses at the boundary are assumed to get dissipated out. As the stress builds up in the system with time, the average stress $\bar{F}$ (average $F_{i,j}$ over all (i,j) cells) increases and the avalanches become bigger and bigger, meaning that any local increase in stress gives rise to more and more nonlocal (distant in both space and time) rearrangements. Eventually, a stationary state is reached with $\bar{F} = F_c$, depending on the model ($F_c \simeq 2.14$ in a square lattice; see Section 1.2.3), when avalanches of all sizes occur (limited only by the size of the system). The energy released in such 'earthquakes' can be identified as the avalanche cluster size, given by the total number of cells toppled by a single initial instability. Figure 4.9 shows the log-log plot of the number $n(S)$ of such earthquakes versus the energy or avalanche size S. The straight line plot indicates the Guttenberg-Richter law for the model: $n(S) \sim S^{-c}$, with

Self-organised criticality and cellular automata models of earthquakes 141

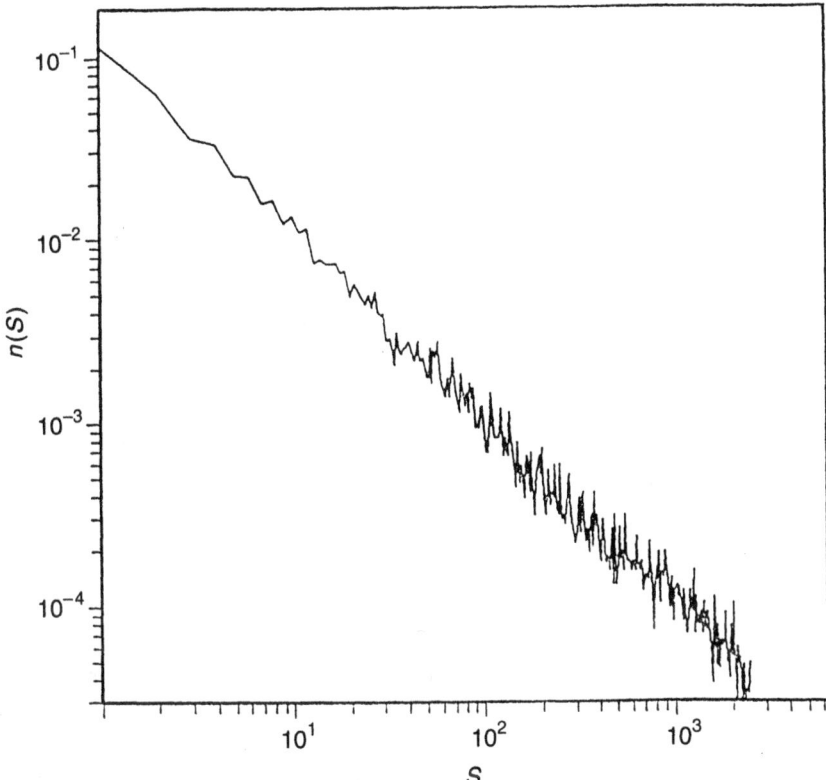

FIG. 4.9. A typical distribution ($n(S)$) of earthquake sizes (S) for a 50×50 BTW automata model, driven (stresses added at randomly chosen sites) at a rate $\sim 2 \times 10^{-5}$ (cf. Bak and Tang 1989).

$c \sim 1.1$. A similar numerical simulation studies for the three dimensional model gives $c \sim 1.3$ (Bak and Tang 1989).

In another, somewhat more realistic automata model, Olami *et al.* (1992) (see Perez *et al.* 1996 for a recent review) considered the mapping of the two-dimensional Burridge-Knopoff spring-block model into a cellular automata model. In fact, if one considers the two-dimensional geometry of the Burridge-Knopoff model as shown in Fig. 4.10, one can write for the total elastic force $F_{i,j}$ on the block at site (i,j), from (4.4),

$$F_{i,j} = \kappa[(2U_{i,j} - U_{i+1,j} - U_{i-1,j}) + (2U_{i,j} - U_{i,j+1} - U_{i,j-1})] + U_{i,j}. \quad (4.7)$$

When this force exceeds the static friction force $\phi(0)$, the block slips. We assume in this model that $F_{i,j}$ becomes zero after a local slip of the block

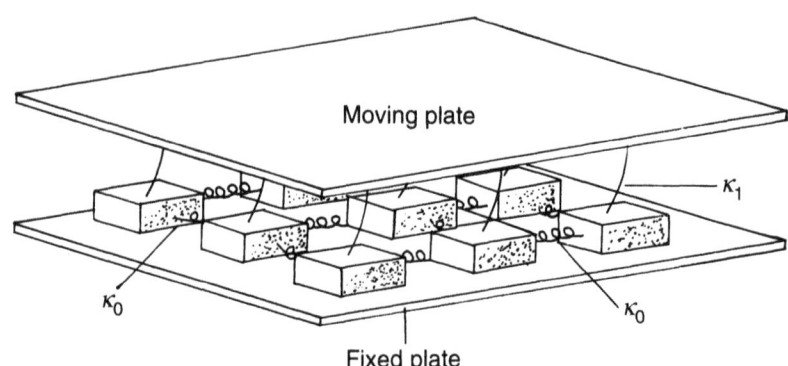

FIG. 4.10. The cellular automata version of the two-dimensional Burridge-Knopoff model, as considered by Olami et al. (1992).

at (i,j). If we denote this additional displacement due to slipping of the block at (i,j), by $\delta U_{i,j}$, then $U_{i,j} \to U_{i,j} + \delta U_{i,j}$, as $F_{i,j} \to 0$. This gives

$$F_{i,j} \to F_{i,j} + (4\kappa + 1)\delta U_{i,j} = 0,$$

and all the forces on the neighbouring blocks of the (i,j)th block get readjusted:

$$F_{i\pm 1,j} \to F_{i\pm 1,j} - \kappa \delta U_{i,j},$$

$$F_{i,j\pm 1} \to F_{i,j\pm 1} - \kappa \delta U_{i,j}.$$

These follow from the contributions of δU_{ij} in $F_{i\pm 1,j}$ and $F_{i,j\pm 1}$, through eqn (4.7). Solving for $\delta U_{i,j}$, we get

$$F_{i\pm 1,j} \to F_{i\pm 1,j} + \delta F_{i\pm 1,j},$$

$$F_{i,j\pm 1} \to F_{i,j\pm 1} + \delta F_{i,j\pm 1},$$

where

$$\delta F_{i\pm 1,j} = \delta F_{i,j\pm 1} = \alpha F_{i,j}, \quad \text{with} \quad \alpha = \frac{\kappa}{4\kappa + 1}. \tag{4.8}$$

Since $\kappa = \kappa_0/\kappa_1$ (in (4.4)) is the ratio of the horizontal and the vertical spring constants of the Burridge-Knopoff model, any nonvanishing value of the vertical spring constant (κ_1) ensures a finite value of κ and $\alpha < 1/4$. This ensures that the total redistributed stress on the neighbours ($\delta F_{i+1,j} + \delta F_{i-1,j} + \delta F_{i,j+1} + \delta F_{i,j-1} = 4\alpha F_{i,j}$) is less than the original stress $F_{i,j}$ at the central block at (i,j), which slips or fails. Thus the stress is not locally conserved during any slip, as assumed in the cellular automata model considered by Bak and coworkers, discussed earlier. Olami et al. (1992) therefore generalised the cellular automata model of earthquake by allowing this local nonconservation of the stress at any event of slip: As $F_{i,j} \geq F_0$, $F_{i,j} \to 0$, and $F_{n.n} \to F_{n.n} + \alpha F_{i,j}$ for the redistribution of forces on the nearest neighbours. Here, the automaton values are also considered as continuous. One can increase F at any randomly chosen cell by a small amount in every iteration, or add $F_0 - F_{\max}$ to all the cells, where $F_{\max}$ is the largest strain anywhere in the previous iteration. One then redistributes the stress around the cell for which the previous stress $F_{\max}$ now becomes F_0, and makes the subsequent redistributions necessary. Repeated operation of this algorithm again brings the system to its self-organised critical state, characterised by avalanches of all sizes. Olami et al. (1992) studied the frequency distribution $n(S)$ of avalanche or earthquake size greater than or equal to S (identified, as before, with the avalanche cluster size), for various values of nonconservation parameter α. They again found the power law $n(S) \sim S^{-c(\alpha)}$ to hold, with the exponent c dependent on the value of α considered. In a simulation for a 35×35 size system, they observed that the experimentally observed value of the exponent c ($0.8 < c < 1.1$) is obtained for a range of values of the nonconservation parameter $\alpha : 0.18 < \alpha < 0.21$ (compared to the stress conserving value of $\alpha = 1/4$ here). This indicates the important role of the spring constants, through nonconservation of local stresses, in determining the Guttenberg-Richter exponent.

4.4 Earthquake fault patterns and percolation model of earthquakes

As mentioned earlier, in the introductory Section 4.1, another aspect of earthquake studies, or seismology in general, is the process in which the flows in the earth's interior mantle strain the outer crusts, driving them in motion and producing intricate patterns of cracks, known as earthquake faults. Indeed these crack or fault patterns show intriguing fractal geometric scaling properties, which some people assume to be indirectly responsible for the Guttenberg-Richter type frequency distribution of the seismic events (Kagan 1982, Barriere and Turcotte 1991, Sahimi et al. 1992). Sahimi et al (1992) analysed a large class of data, collected by various groups and agencies, for the two-dimensional cross-section of the crack pattern of the rocks near several geothermal geyser fields in central and northern California (see e.g. Fig. 4.11), earthquake hypocentres in various parts of Japan, etc. They found that the cracked pavements have a fractal structure, with the number n of cracks within a box of size L growing as $n \sim L^{d_f}$, with d_f in the range $1.6 \leq d_f \leq 1.9$, from the fitting of the data in the sample size

FIG. 4.11. A typical cross-section of the earthquake fault pattern in the rocks near a geysers field (from Sahimi et al. 1992).

range one to a hundred kilometres. This fractal dimension d_f corresponds to the entire cluster of the percolating cracks. It was found that in the smaller length scales, in the range one to about ten kilometres, a major fault is not a straight line crack, but rather a fractal with the fractal dimension $d_B \sim 1.2$. Sahimi et al. identified both d_f and d_B as the dimensions of the percolation cluster and of the backbone of the fractured network of random composites, discussed in Section 3.6 of the previous chapter, where the threshold strength distribution of each bond is distributed randomly with a power law. Two-dimensional simulations of such systems indeed gave very similar values ($d_f \sim 1.7$ and $d_B \sim 1.2$) for the percolation cluster size scaling exponent and the infinite backbone size scaling exponent (see Section 3.6.2). These results and the success in comparisons also indicate that in order to transmit the stress and produce the consequent deformation, the earthquake hypocentres must be situated on the backbone of the network of faults, and hence the geometric scaling properties of the structure of the earthquake faults near the hypocentres. It is difficult, at this stage, to extend the implications of this power law scaling of the geometry of earthquake faults to the power law scaling of the frequency of the earthquake energy releases. However, there has been conjecture (see e.g. Sahimi et al. 1992) that the exponent c in the Guttenberg-Richter law $n(E_r) \sim E_r^{-c}$ can be identified as the fractal dimension of the backbone of the earthquake of the faults: $c = d_B \simeq 1.2$, as most of the earthquake dynamics occurs in the crust ($d = 2$). As mentioned earlier, no real justification for connecting the above two quantities exists, and hence the proposal may be accepted with proper caution.

4.5 Response of sandpile models to weak pulses and precursors of self-organised criticality and earthquakes

As we have seen in the previous section, earthquakes are identified with avalanches in the self-organised critical state. The power law distribution of the avalanche sizes in this (self-organised) critical state is identified with the Guttenberg-Richter law. We have also seen, in the Burridge-Knopoff model, that after the charging time is over, the strain energy of the system attains a 'critical' state on the verge of chaos, for appropriate ranges of values of the parameters of the dynamics. The excess elastic energy is then released in such a critical state through a random series of slips, having a power law distribution. By this (slipping) process, due to the rather large slips or earthquakes, the system often becomes 'sub-critical', and gets charged again to the critical state due to its continuous (sticking) dynamics.

Obviously, it would be quite useful to know the precursors of such critical states, and be able to predict imminent catastrophes. It has been suggested recently (by Acharyya and Chakrabarti 1996a,b) that looking at the growth of the responses to the appropriate local and weak pulsed perturbations in some models, the approach to the self-organised critical state can be studied and its appearance can be predicted. This has been demonstrated, in particular, in the BTW critical height model (introduced in Section 1.2.3 of the first chapter, and also discussed in the previous section). It has been shown that by adding a fixed small amount of 'sand grains' or heights locally at any lattice site for a fixed and small time period δt to several realisations of the BTW system with various average heights $\bar{z} < z_c$, one can measure the spread of the 'damage' by counting the number of sites toppled around the local perturbation site to adjust the instability, and also measuring the response time $\Delta t(\bar{z})$ to absorb the instabilities induced by the pulsed perturbation, by counting the number of iterations required for stabilisation of the dynamics (when the topplings get stopped). It has been shown that the ratio $R(\bar{z}) = \Delta t/\delta t$ of the response time to the time width of the pulsed perturbation diverges following a power law $R(\bar{z}) \sim (\bar{z} - z_c)^{-\gamma}$. This suggests that a plot of $R^{-1/\gamma}$ against $\bar{z}$ would give a straight line extrapolating to the self-organised critical point z_c where $R^{-1/\gamma}$ vanishes. One can thus predict the critical point z_c by measuring the responses to very weak pulsed perturbations (by adding small amounts of 'sands' locally in space and time) in the subcritical states far away from the self-organised critical state (with $\bar{z} < z_c$). It may be mentioned here that the pulsed addition (of 'sands') being local in both space and time, the perturbation does not affect in practice the total average height $\bar{z}$. The measurement of the local damage spread also shows growth (approaching a finite peak depending on the pulse width δt), and can be utilised similarly to predict the self-organised critical point, from the measurements in the states far away from the criticality. One can make a similar prediction about the value of the global failure or earthquake stress F_c in the cellular automata model discussed in Section 4.3, by appropriately redefining the quantities. Similar studies have also been made for the Burridge-Knopoff model of earthquakes (Acharyya and Chakrabarti 1995, 1996a,b).

In the following, we give some details of these studies of response to weak pulses in the BTW model, and of weak periodic pulses in the Burridge-Knopoff model.

4.5.1 *Pulse response of the sandpile model*

Let us consider the self-organised criticality of the critical height BTW model introduced in Section 1.2.3. Here, one considers finite lattices, with absorbing boundaries, of sizes L^2 on square and L^3 on simple cubic lattices. At each lattice point the 'heights' or 'sand grains' or 'stresses' are randomly added in discrete integer addition and avalanches take place if the height z_i at any point i exceeds the value 3 in square and 5 in simple cubic lattices respectively (the cut-off values $z_0 = 4$ and 6 respectively). In such cases, the $z_{i\pm\delta}$ of the nearest neighbours δ of the site i gets one unit of height each and z_i becomes zero at i. The dynamics continues, until all the sites have $z < z_0$. The simulation studies give the value of the average (self-organised) critical height z_c to be around 2.124 in square and 2.654 in simple cubic lattices respectively for such a model, beyond which the global avalanches take place (see Section 1.2.3 of Chapter 1).

We now study the effect of addition of a fixed (small) number h_p of sand grains or stresses or heights at any central point for a time period δt, when the system has the average height $\bar{z}$ ($< z_c$) and the dynamics (before the addition of these particles) has stopped. Immediately after the particles are added, the local dynamics starts again and it continues, for a time period $\Delta t(\bar{z})$ ($\geq \delta t$). One now measures the ratio $R(\bar{z}) = \Delta t/\delta t$ of the response time to the perturbation (pulse) time. From the simulation results (Acharyya and Chakrabarti 1996a,b) for $L = 150$ in square lattices and 20 in simple cubic lattices, one observes that $R \sim (z_c - \bar{z})^{-\gamma}$, where $\gamma \cong 1/3$ in both $d = 2$ and 3. One can thus clearly locate the self-organised critical point (or critical 'height') z_c by plotting $R^{-1/\gamma}$ with $\bar{z}$ and by locating, from straight line extrapolation, its vanishing point (where R diverges). This gives $z_c \cong 2.16$ in the case of square lattices and $z_c \cong 2.66$ in the case of simple cubic lattices. These estimated values of z_c are indeed very close to the previous straightforward numerical estimate $z_c \cong 2.124$ and $\cong 2.654$ in square and simple cubic lattices respectively (Grassberger and Manna 1990). It may be mentioned that the above estimate of the value of the exponent γ in this model is, so far, tentative and simulations of much larger size lattices (with $L \sim 1000$) indicate the value of γ to be of the order of unity (Acharyya and Chakrabarti 1996b). It may also be mentioned that a similar critical slowing-down behaviour, for the BTW system approaching its SOC state ($\bar{z} \to z_c$), was indeed anticipated by Grassberger and Manna (1990), where of course the attempt was abandoned since $\bar{z}$ was observed not to describe completely the SOC states (because of overshooting of $\bar{z}$ over its critical value z_c as the system is driven towards the critical point).

Like the growth of response time (Δt, for fixed perturbation time δt), with approach to criticality ($\bar{z} \to z_c$) in the BTW model (as discussed above), the size (length) of the 'damaged' region (for the additional pulse of sand or height at a fixed site) also increases as $\bar{z} \to z_c$. To have a quantitative measurement, one can count the number $\Delta n(\bar{z})$ of sites affected (toppled at least once) due to a fixed addition of sand (height) at a particular (central) point in the BTW model. This $\Delta n(\bar{z})$ also grows as the average height $\bar{z}$ approaches the critical point $\bar{z}_c$. However, Δn does not diverge for any finite pulse width δt. Rather, Δn grows upto a peak value dependent on δt as $\bar{z} \to z_c$. Because of the nonuniversal value of this peak, the study of the growth of $\Delta n(\bar{z})$ is not very convenient for predicting the value of z_c.

4.5.2 *Response of the Burridge-Knopoff model to localised periodic pulses*

As discussed in Section 4.2, in the Burridge-Knopoff dynamical model of earthquakes, the elastic stresses (energy) developed compete with the (velocity-dependent) frictional force as long as the blocks stick to the moving platform. As it fails, there occurs local failure (slippage for a finite number of blocks) or global failures (simultaneous slippage of almost all the blocks). Initially, during the charging period, the (elastic) energy increases and reaches a critical state characterised by slips of all sizes. The eventual large slips, involving almost all the blocks, often release elastic energy more than the 'excess', and the system again becomes 'sub-critical'. The energy starts to build up again and after some (recharging) time it becomes critical again and quakes of all sizes occur; and the process continues.

Like the approach to the self-organised critical state in the BTW model, as discussed in the previous section, one can study the approach to the 'critical' state in the Burridge-Knopoff model, by looking at the response to suitable weak pulses periodically, with the periodicity much less than the typical charging period of the system. In a recent numerical study (Acharyya and Chakrabarti 1995, 1996a,b), of the solution of the Burridge-Knopoff system (4.4), additional weak pulses of fixed magnitude and duration were periodically applied to a central block. This process imparted regularly a fixed small amount of momentum to that (arbitarily chosen) block periodically. From the maximum displacement of the block within the time between two successive pulses, one can estimate the sensitivity of the response of the system. As the system approaches the crtical point (due to its own dynamics caused by the relative motion of the platform), this sensitivity in response magnitude or the 'susceptibility' is observed to grow and assumes a (nonuniversal) value depending on the dynamical parameters. Studying the growth of the response (or susceptibility) to such fixed amplitude (and width) periodic pulses, one can estimate the imminent global avalanches or quakes.

Stated briefly, it appears that one can define appropriate susceptibilities for systems having macroscopic breakdown properties. As the (global) breakdown point approaches (for example, with the increase of time as in the BTW model of sandpiles or in the Burridge-Knopoff model of earthquakes), the appropriate correlations grow and the corresponding susceptibility grows and assumes a peak value (or diverges) at the disaster point. By investigating the response to pulsed perturbations in such systems, one can estimate this susceptibility and by locating the extrapolated point where its inverse (with some appropriate power) decreases or vanishes, one can make predictions about the imminent breakdown point.

It may be mentioned here in connection with the studies on the precursor effects of earthquakes and their prediction, Bufe and Varnes (1993) and Sornette and Sammis (1995) observed that the normal seismic activity (without any pulsed perturbations etc.), measured by the strain release $\epsilon_r(t)$ before the earthquake, seems to grow as a power of the time interval $(t_q - t)$ from the next major earthquake occurring at time t_q, with log-periodic corrections:

$$\epsilon_r(t) = A + B(t_q - t)^{\tilde{m}}[1 + C\cos D\ln(t_q - t) + E], \qquad (4.9)$$

where A, B, C, D and E are constants and $\tilde{m}$ is a critical exponent. In fact, the fit (by Bufe and Varnes 1993) to eqn (4.9) of the accelerating seismic activity of the Loma Prieta earthquakes in north California, prior to the year 1989, gives a fitting value of $t_q = 1989\pm0.8$, while an actual earthquake of magnitude 6.7-7.1 occurred in October 1989. Sornette and Sammis (1995) attempted to provide renormalisation group theoretic justifications for the above equation. As mentioned in an earlier chapter (Section 3.7.3), the growth of the elastic energy release with strain in heterogeneous elastic networks also follows a very similar power law (Sahimi and Arbabi 1996).

4.6 Summary and conclusions

As one can see, not much has been understood yet about the earthquake mechanism and its dynamics. The Guttenberg-Richter power law (4.1) distribution of the earthquake magnitudes is, so far, the only well formulated observation to guide us to model such failure phenomena. As mentioned before, the Guttenberg-Richter law is considered to be a strong indication of a critical state of the earthquake dynamics, achieved of course in a self-organised way. Most of the theoretical models discussed here indeed attempt to capture such a property for the dynamics of failures in the respective models. Although some precursors for the self-organised critical state can be seen (see Section 4.5) in the cellular automata models, their implications for the other earthquake models are not clear yet.

The role of disorder, in particular of the fractal structure of the earthquake faults (discussed in Section 4.4), are not clearly understood. As discussed in an earlier chapter (Section 3.8), the dynamics of fracture in disordered solids also indicate similar (Guttenberg-Richter type) power law behaviour in the power spectrum of the ultrasonic emission from such solids, as the fracture propagates. No doubt the understanding of the connections between the dynamics of fracture in disordered solids and the dynamics of earthquakes will become much clearer in the near future, because of the intensive efforts which are being made currently.

REFERENCES

Abraham, F. F. (1996). *Physical Review Letters*, **77**, 869
Abraham, F. F., Brodbeck, D., Rafey, R. and Rudge, W.E. (1994). *Physical Review Letters*, **73**, 272
Acharyya, M. and Chakrabarti, B. K. (1995). *Indian Journal of Physics A*, **69**, 205
Acharyya, M. and Chakrabarti, B. K. (1996a). *Physical Review E*, **53**, 140
Acharyya, M. and Chakrabarti, B. K. (1996b). *Physica A*, **224**, 254
Acharyya, M. Ray, P. and Chakrabarti, B. K. (1996). *Physica A*, **224**, 287
Allen, M. P. and Tildesley, D. J. (1987). *Computer Simulation of Liquids*, Clarendon Press, Oxford
Ananthakrishna G., and Ramachandan, H. (1994). In *Nonlinearity and Breakdown in Soft Condensed Matter*, ed. K. K. Bardhan, B. K. Chakrabarti and A. Hansen, Lecture Notes in Physics, volume 437, Springer-Verlag, Heidelberg, p. 78
Bak, P. and Tang, C. (1989). *Journal of Geophysical Research*, **94**, 15635
Bak, P., Tang, C. and Weisenfeld, K. (1987). *Physical Review Letters*, **59**, 381
Bak, P., Tang, C. and Weisenfeld, K. (1988). *Physical Review A*, **38**, 364
Barabasi and Stanley, H. E. (1995). *Fractal Concepts in Surface Growth*, Cambridge University Press, Cambridge.
Bardhan, K. K. and Chakrabarti, R. K. (1994). *Physical Review Letters*, **72**, 1068
Bardhan, K. K., Chakrabarti, B. K. and Hansen, A. (ed.) (1994). *Nonlinearity and Breakdown in Soft Condensed Matter*, Lecture Notes in Physics, Volume 437, Springer, Heidelberg.
Barriere, B. and Turcotte, D.L. (1991). *Geophysical Research Letters*, **18**, 2011.
Beale P. D. and Duxbury, P. M. (1988). *Physical Review B*, **37**, 2785
Beale P. D. and Srolovitz D. S. (1988). *Physical Revew B*, **37**, 5500
Benguigui, L. (1988). *Physical Review B*, **38**, 7211
Benguigui, L. and Ron, P. (1993). *Physical Review Letters*, **70**, 2423
Benguigui, L. and Ron, P. (1994). In *Nonlinearity and Breakdown in Soft Condensed Matter*, ed. K. K. Bardhan, B. K. Chakrabarti and A. Hansen, Lecture Notes in Physics, Vol. 437, Springer, Heidelberg, p. 221
Benguigui, L., Ron P. and Bergman, D. J. (1987). *Journal de Physique*, **48**, 1547

Bergman, D. J. (1985). *Physical Review B*, **31**, 1696

Bergman, D. J. (1986). In *Fragmentation, Form and Flow in Fractured Media*, ed. R. Englman and J. Jeager, Annals of Israel Physical Society, Vol. 8, p. 266

Bergman, D. J. and Kantor, Y. (1984). *Physical Review Letters*, **53**, 511

Bergman, D. J. and Stroud, D. (1992). In *Solid State Physics*, Vol. **46**, ed. H. Ehrenreich, and D. Turnbull, Academic Press, N. Y., p. 147

Born, M. and Huang, K. (1956). *Dynamical Theory of Crystal Lattices*, Clarendon Press, Oxford

Bouchaud, E. and Bouchaud, J. P. (1994). *Physical Review B*, **50**, 17752

Bouchaud, J. P., Bouchaud, E., Lapasset G. and Planes, J. (1993a). *Physical Review Letters*, **71**, 2240

Bouchaud, E., Lapasset, G., Planes, J. and Naveos, S. (1993b). *Physical Review B*, **48**, 2917

Bowden, F. P., and Tabor, D. (1950). *Friction and Lubrication of Solids*, Clarendon Press, Oxford

Bowman D. R. and Stroud, D. (1985). *Bulletin of the American Physical Society*, **30**, 563

Bowman, D. R. and Stroud, D. (1989). *Physical Review B*, **40**, 4641

Brace, W. F. and Orange, A. S. (1988). *Journal of Geophysical Research*, **73**, 1433

Bradley, R. M., and Wu, K. (1994). *Physical Review E*, **50**, R631

Brechet, Y., Bellet, D. and Neda, Z. (1993). In *Nonlinear Phenomena in Material Science III: Instabilities and Patterning*, ed. G. Ananthakrishana, L. P. Kubin and G. Martin, Sci-Tech Publications, Switzerland, p. 247

Bufe, C. G. and Varnes, D. J. (1993). *Journal of Geophysical Research*, **98**, 9871

Bunde, A., Coniglio, A., Hong, D. C. and Stanley, H. E. (1985). *Journal of Physics A: Mathematical and General*, **18**, L137

Burridge, R., and Knopoff, L. (1967). *Bulletin of the Seismological Society of America*, **57**, 341

Caldarelli, G., Di Tolla, F. D. and Petri, A. (1996). *Physical Review Letters*, **77**, 2503

Carlson, J. M., and Langer, J. S. (1989a). *Physical Review Letters*, **62**, 2632

Carlson, J. M., and Langer, J. S. (1989b). *Physical Review A*, **40**, 6470

Carlson, J. M., Langer, J. S., and Shaw, B. E. (1994). *Reviews of Modern Physics*, **66**, 657

Chakrabarti, B. K. (1988). *Review of Solid State Science*, **2**, 559

Chakrabarti, B. K. (1994). In *Nonlinearity and Breakdown in Soft Condensed Matter*, ed. K. K. Bardhan, B. K. Chakrabarti and A. Hansen,

Lecture Notes in Physics, Vol. 437, Springer, Heidelberg, p. 171
Chakrabarti, B. K., Chowdhury, D. and Stauffer, D. (1986). *Zeitschrift fur Physik B*, **62**, 343
Chakrabarti, B. K., Roy, A. K. and Manna, S. S. (1988). *Journal of Physics C: Solid State Physics*, **21**, L65
Charmet, J. C., Roux, S. and Guyon, E. (eds) (1990). *Disorder and Fracture*, Plenum Press, New York and London
Chayes, J. T., Chayes, L. and Durret, R. (1986). *Journal of Statistical Physics*, **45**, 933
Coppard, R. W., Dissado, L. A., Rowland, S. M. and Rakowski, R. (1989). *Journal of Physics C: Condensed Matter*, **1**, 3041
De Arcangelis, L. (1990). In *Statistical Models for the Fracture of Disordered Media*, ed. H. J. Herrmann and S. Roux, North Holland, p. 229
De Arcangelis, L., Redner, S. and Herrmann, H. J. (1985). *Journal de Physique Letters*, **46**, 585
De Gennes, P. G. (1976). *Journal de Physique Letters*, **37**, L1
Dhar, D. (1990). *Physical Review Letters*, **64**, 1613
Dienes, G. J. and Paskin, A., (1983) In *Atomistics of Fracture,* ed. R. M. Latanison, and J. R. Pickens, Plenum Press, N.Y., p. 671
Duke, C. B. (1969). In *Solid State Physics*, Suppl. Vol. 10, ed. F. Seitz, D. Turnbull and H. Ehrenreich, Academic Press, N. Y.
Duxbury, P. M. (1990). in *Statistical Methods for the Fracture of Disordered Media*, ed. H. J. Herrmann and S. Roux, North Holland, Amsterdam, p. 189
Duxbury, P. M. and Leath, P. L. (1987). *Journal of Physics A*, **20**, L411
Duxbury, P. M. and Li, Y. (1990). In *Disorder and Fracture*, ed. C. J. Charmet, S. Roux and E. Guyon, Plenum Press, N. Y., p. 141
Duxbury, P. M., Beale, P. D. and Leath, L. P. (1986). *Physical Review Letters*, **57**, 1052
Duxbury, P. M., Leath, P. L. and Beale, P. D. (1987). *Physical Review B* **36** 367
Englman, R. and Jeager, J. (ed.) (1986). *Fragmentation, Form and Flow in Fractured Media*, Annals of Israel Physical Society, Vol. 8
Evans, A. G. and Zok, F. W. (1986). In *Solid State Physics*, Vol. 47, ed. H. Ehrenreich and D. Turnbull, Academic Press, N. Y., p. 177
Garfunkel, G. A. and Weissman, M. B. (1985). *Physical Review Letters*, **55**, 296
Gefen, Y., Shih, W. H., Laibowitz, R. W. and Viggiano, J. M. (1986). *Physical Review Letters*, **57**, 3097
Gilabert, A., Vaneste, C. and Sornette, D. (1987). *Journal de Physique*, **48**, 763
Grassberger, P. and Manna, S. S. (1990). *Journal de Physique*, **51**, 1091

Griffith, A. A. (1920). *Philosophical Transactions of the Royal Society, London A*, **221**, 163
Gumbel, E. J. (1958). *Statistics of Extremes*, Columbia University Press, New York
Guttenberg, B. and Richter, C. F. (1954). *Seismicity of the Earth and Associated Phenomena*, Princeton University Press, Princeton, N. J.
Halperin, B. I., Feng, S. and Sen, P. N. (1985). *Physical Review Letters*, **54**, 2391
Halpin-Healy, T. and Zhang, Y. C. (1995). *Physics Reports*, **254**, 215.
Hansen, A., Hinrichsen, E. L. and Roux, S. (1990). *Physica Scripta*, **T33**, 20
Hansen, A., Hinrichsen, E. L. and Roux, S. (1991). *Physical Review B*, **43**, 665
Herrmann, H. J. (1990). *Pyhsica A*, **163**, 359.
Herrmann, H. J. and Roux, S. (eds) (1990). *Statistical Models for the Fracture of Disordered Media*, North-Holland, Amsterdam
Herrmann, H. J., Hansen, A. and Roux, S. (1989). *Physical Review B*, **39**, 637
Heslot, B., Baumberger, T., Perrin, B., Caroli, B. and Caroli, C. (1994). *Physical Review E*, **49**, 4973
Huntington, H. B. (1975). In *Diffusion in Solids − Recent Developments*, ed. A. S. Novick and J. J. Burton, Academic Press, N. Y.
Inglis, C. E. (1913). *Transactions of the Institute of Naval Architecture*, **55**, 219
Jacobs, D. J., and Thorpe, M. F. (1995) *Physical Review Letters*, **75**, 4051
Jagota, A. and Bennison, S. J. (1994). In *Nonlinearity and Breakdown in Soft Condensed Matter*, ed. K. K. Bardhan, B. K. Chakrabarti and A. Hansen, Lecture Notes in Physics, Volume 437, Springer, Heidelberg, p. 186
Jayatilaka, A. de S. (1979). *Fracture of Engineering Brittle Materials*, Applied Science Publishers, London
Kagan, Y. Y. (1982). *Geophysical Journal of Royal Astrophysical Society of Canada*, **84**, 2348
Kahng, B., Batrouni, G. G., Redner, S., de Arcangelis, L and Herrmann, H. J. (1988). *Physical Review B*, **37**, 7625
Kardar, M., Parisi, G. and Zhang, Y. C. (1986). *Physical Review Letters*, **56**, 889
Kar-Gupta, A. and Sen, A. K. (1995). *Physica A*, **215**, 1
Kertesz, J. (1992). *Physica A*, **191**, 208.
Kertesz, J., Horrath, V. K. and Weber, F. (1993). *Fractals*, **1**, 67
Kostrov, B. V., and Das, S. (1988). *Principles of Earthquake Source Mechanics*, Cambridge University Press, Cambridge

Lamaignere, L., Carmona, F. and Sornette, D. (1996). *Physical Review Letters*, **77**, 2738

Langer, J. S. (1993). *Physical Review Letters*, **70**, 3592

Langer, J. S. and Nakanishi, H. (1993). *Physical Review E*, **48**, 439

Larralde, H. and Ball, R. C. (1995). *Europhysics Letters*, **30**, 87

Lawn, B. R., and Wilshaw T. (1975). *Fracture of Brittle Solids*, Cambridge University Press, Cambridge

Leath, P. L. and Duxbury, P. M. (1994). In *Nonlinearity and Breakdown in Soft Condensed Matter*, ed. K. K. Bardhan, B. K. Chakrabarti and A. Hansen, Lecture Notes in Physics, vol. 437, Springer, Heidelberg, p. 151

Li, Y. S. and Duxbury, P. M. (1987). *Physical Review B*, **36**, 5411

Lobb, C. J., Hui, P. M., and Stroud, D. (1987). *Physical Review B*, **36**, 1956

Maloy, K. J., Hansen, A. Hinrichsen, E. L. and Roux, S. (1992). *Physical Review Letters*, **68**, 213

Mandelbrot, B. B., Passoja, D. E. and Pullay, A. J. (1984). *Nature*, **308**, 721

Manna, S. S. (1991). *Journal of Physics A*, **24**, L363

Manna, S. S. and Chakrabarti, B. K. (1987). *Physical Review B*, **36**, 4078

Marder, M. and Fineberg, J. (1996). *Physics Today*, September, 24

Marder, M. and Liu, X. (1993). *Physical Review Letters*, **71**, 2417

Meakin P. (1990). In *Statistical Models for the Fracture of Disordered Media*, ed. J. J. Herrmann and S. Roux, North Holland, Amsterdam, p. 291

Mecholsky, J. J., Makin, T. J. and Pasoja, D. E. (1988). *Advances in Ceramics*, **22**, 127

Mecholsky, J. J., Pasoja, D. E. and Feinberg-Ringel, R. S. (1989). *Journal of the American Ceramic Society*, **72**, 60

Mendelson, K. S. (1975). *Journal of Applied Physics*, **46**, 917, 4740

Morse, P. M. and Feshbach, H. (1953). *Methods of Theoretical Physics*, Volumes I & II, McGraw-Hill, New York

Mosolov, A. B. (1993). *Europhysics Letters*, **24**, 673

Mott, N. F. (1948). *Engineering*, January, 16

Moukarzel, C., and Duxbury, P. M. (1995). *Physical Review Letters*, **75**, 4055

Myers, C. R., Shaw, B. E. and Langer, J. S. (1996). *Physical Review Letters*, **77**, 972

Nakano, A., Kalia, R. K. and Vashistha, P. (1995). *Physical Review Letters*, **75**, 3138

O'Dwyer, J. (1973). *The Theory of Electrical Conduction and Breakdown in Solid Dielectrics*, Clarendon Press, Oxford

Olami, Z., Feder, H. J. S. and Christensen, K. (1992). *Physical Review Letters*, **68**, 1244

Paskin, A., Gohar, A. and Dienes, G. J. (1980). *Physical Review Letters*, **44**, 940

Paskin, A., Som, D. K. and Dienes, G. J. (1981). *Journal of Physics C: Solid State Physics*, **14**, L171

Perez, C., Corral, A., Diaz-Guilera, A., Christensen, K., and Arenas, A. (1996). *International Journal of Modern Physics B*, **10**, 1111

Petri, A., Paparo, G., Vespignani, A., Alippi, A. and Costantini, M. (1994). *Physical Review Letters*, **73**, 3423

Rammal, R., Tannous, C., Benton, P. and Trembley, A. M. S. (1985a). *Physical Review Letters*, **54**, 1718

Rammal R. , Tannous C. and Trembley, A. M. S. (1985b). *Physical Review A*, **31**, 2662

Rautiainen, T. T., Alva, M. J. and Kaski, K. (1995). *Physical Review E*, **51**, R2727

Ray, P. and Chakrabarti, B. K. (1985a). *Solid State Communications*, **53**, 477

Ray, P. and Chakrabarti, B. K. (1985b). *Journal of Physics C*, **18**, L185

Ray, P. and Chakrabarti, B. K. (1988). *Physical Review B*, **38**, 715

Ray, P. and Date, G. (1996). *Physica A*, **229**, 26

Rosen, W. R. and Mamun, C. K. (1993). *Physical Review*, **47**, 11815

Roux, S. (1994). In *Nonlinearity and Breakdown in Soft Condensed Matter*, Lecture Notes in Physics, Vol. 437, ed. K. K. Bardhan, B. K. Chakrabarti and A. Hansen, Springer Verlag, Heidelberg, p. 235

Roux, S., and Herrmann, H. J. (1987). *Europhysics Letters*, **4**, 1227

Roux, S., Hansen, A. and Guyon, E. (1987) *Journal de Physique* , **48**, 2125

Sahimi, M. (1992). *Application of Percolation Theory*, Taylor & Francis, London

Sahimi, M. (1995). In *Computational Physics*, Vol. II, ed. D. Stauffer, World Scientific, Singapore, p. 175

Sahimi, M. (1997). *Physics Reports*, (in press)

Sahimi, M. and Arbabi, S. (1992). *Physical Review Letters*, **68**, 608

Sahimi, M. and Arbabi, S. (1993). *Physical Review B*, **47**, 713

Sahimi, M. and Arbabi, S. (1996). *Physical Review Letters*, **77**, 3689

Sahimi, M. and Goddard, J. D. (1986). *Physical Review B*, **33**, 7848

Sahimi, M., Robertson, M. and Sammis, C. G. (1992). *Physica A*, **191**, 57

Schmittbhul, J., Vilotte, J. P. and Roux, S. (1993). *Europhysics Letters*, **21**, 375

Scholz, C. H. (1990). *The Mechanics of Earthquakes and Faulting*, Cambridge University Press, N. Y.

Sen, A. K. and Kar-Gupta, A. (1994). In *Nonlinearity and Breakdown in Soft Condensed Matter*, ed. K. K. Bardhan, B. K. Chakrabarti, and H. Hansen, Lecture Notes in Physics, Volume 437, Springer-Verlag, Heidelberg, p. 271
Sen, P. (1996). *International Journal of Modern Physics C*, **7**, 603
Sieradzki, K. and Li, R. (1986). *Physical Review Letters*, **56**, 2509.
Skal, A. and Shklovskii, B. I. (1974). *Soviet Physics: Semiconductors*, **8**, 1029
Sornette, D. (1987). *Journal de Physique*, **48**, 1843
Sornette, D. and Sammis, C. G. (1995). *Journal de Physique I*, **5**, 607
Sornette, D. and Vanneste, C. (1992). *Physical Review Letters*, **68**, 612
Stanley, H. E. (1977). *Journal of Physics A: Mathematical and General*, **10**, L211
Stauffer, D. (1979). *Physics Reports*, **54**, 1
Stauffer, D. and Aharony, A. (1992). *Introduction to Percolation Theory*, Taylor & Francis, London
Stinchcombe, R. B., Duxbury, P. M. and Shukla, P. K. (1986). *Journal of Physics A: Mathematical and General*, **19**, 3903
Thomson, R. (1986). In *Solid State Physics*, Vol. 39, ed. H. Ehrenreich and D. Turnbull, Academic Press, N.Y., p. 1
Van den Born, I. C., Santen, A., Hoekstra, H. D. and de Hosson, J. Th. M. (1991). *Physical Review B*, **43**, 3794
Vasconcelos, G. L. (1996). *Physical Review Letters*, **76**, 4865
Vilotte, J. -P., Schmittbhul, J., and Roux, S. (1994). In *Nonlinearity and Breakdown in Soft Condensed Matter*, ed. K. K. Bardhan, B. K. Chakrabarti and A. Hansen, Lecture Notes in Physics, Volume 437, Springer-Verlag, Heidelberg, p. 54
Weibull, W. (1951). *Journal of Applied Mechanics*, **18**, 293
Wright, D. C., Bergman, D. J. and Kantor, Y. (1986). *Physical Review B*, **33**, 396
Wu, K. and Bradley, R. M. (1994). *Physical Review B*, **50**, 12468
Yagil, Y. (1992). Unpublished thesis, Tel-Aviv University
Yagil, Y. and Deutscher, G. (1992). *Physical Review B*, **46**, 16115
Yagil, Y., Deutscher, G. and Bergman, D. J. (1992). *Physical Review Letters*, **69**, 1423
Yagil, Y., Deutscher, G. and Bergman, D. J. (1993). *International Journal of Modern Physics B*, **7**, 3353
Yuse, A. and Sano, M. (1993). *Nature*, **362**, 329
Zabolitzky, J. G., Bergman, D. J. and Stauffer D. (1986). *Journal of Statistical Physics*, **44**, 211
Zapperi, S., Ray, P., Stanley, H. E. and Vespignani, A. (1997). *Physical Review Letters*, **78**, 1408.

Zhang, S. Z., Lung, C. W. and Wang, K. L. (1990). *Physical Review B*, **42**, 6631

INDEX

AC field 56, 57
aftershock relaxation 126
angle of repose 27
annealed impurity/disorder 126
avalanches 28, 29, 79, 140, 145

backbone dimension 12, 145
backbone of percolation cluster 11–13
Bethe lattice 14, 15
blobs of percolation cluster 11
bond-bending elastic network 15, 104, 110, 114
bond disorder 7, 17, 75, 76
breakdown model statistics 1, 3, 22–27, 37–41, 65, 66, 106–113
breakdown susceptibilities 73–75, 79, 148
brittle solids 1, 4, 49, 83, 105, 106
BTW model 28, 140, 146–148
Burridge-Knopoff model 4, 130–139, 148

cellular automata model 28, 29, 130, 140–143
  earthquake 140–143
  sandpile 28, 29, 130, 140, 146–148
central force percolation 17, 91, 104

chaotic motion 120, 121, 136, 140
chemical length 12, 13
clusters 7, 8, 9, 11, 23, 24
computer simulation 77, 89–91, 109, 110, 119–121
conjugate gradient method 103, 104
connectivity 8–10
continuum percolation 18–20, 42, 43, 67
correlation length 9, 13, 24
crack 20, 23, 82–95
  growth 117–125
  propagation 117–125
  surface 92–95, 119–121
  tip 21, 22, 87, 121
critical exponents 2, 8–20
critical slowing down 119
cumulative failure probablity 23–25, 40

DC field 1, 56
defects 20–24, 33–35, 39, 43, 44, 65, 66
  dangerous/ weakest/ worst 23, 24, 35, 38, 39, 43, 44, 66

elliptic 20–22, 35, 85
dielectric breakdown 1, 30, 62–78
dimensionality
  fractal 12–14, 117, 144, 145
  lattice 9, 36, 45, 67, 96–98
distribution of failure threshold 48–52, 74–76
double-exponential form 5, 24, 40, 65, 73, 109
duality (transformation) 15, 62–64
ductile solids 33, 83, 100, 49
dynamics
  of breakdown 73, 74, 79
  of fracture 117–127

earthquake
  faults (pattern) 143–145
  models 130–143
  statistics 28, 29, 128–130
elastic energy
  released in earthquake 129, 131–133
  released in fracture 86, 118, 124–126
elasticity 1, 15–16
elastic moduli 2, 16, 22
elastic network
  bond-bending 15, 16, 104, 110
  central force 17, 91, 104
electromigration failure 52–56
elliptic coordinates 20, 21
elliptic hole or defect 20–22, 35, 85
experimental results 48, 59, 60, 76–78, 94, 95, 100, 101, 110–113, 125, 126
experiments 76–78, 88, 89, 94, 95, 98, 99, 130–133
exponents 2, 8–20, 41–45, 65–67
  conductivity 2, 14
  correlation length 9
  dielectric breakdown 67
  elasticity 2, 16, 17
  fractals 12–14
extreme statistics 22–27, 37–41, 65, 66, 106–113

failure
  statistics 22–27, 38–40, 65–67, 106–109, 116, 117

strength (stress) 26, 86–88, 107
  voltage 32, 64–68, 77, 78
fault pattern (earthquake) 130, 143–145
final breakdown
  current 79
  field 74, 79
  strength/stress 121–125
fluctuations 26, 72
fractals 12–14, 117, 143–145
  dimensions 12, 13, 14, 117
  percolating 12
  Sierpinsky carpet 13
fracture
  growth 117–127
  propagation 117-122
  roughness 91–95, 119–122
  roughness exponent 91–95
  strength 86–105
  surface 91–95, 119–122
friction 130, 133–135
frictional instability 133–135
fuse model 1, 30, 33–61

geology 129
geothermal geysers 143
global failure 28, 146, 147, 148
Griffith's law 24, 82–95
growth
  of breakdown 73, 74 79
  of fracture 117–127
  of fuse 79
Gumbel statistics/distribution 3, 24–26, 37–41, 65, 107–113, 117
Guttenberg-Richter law 4, 28, 79, 126, 128–130

infinite cluster 11
initial breakdown
  current 79
  field 74, 79
  stress 121–125
instabilities 3, 49, 50, 91, 120, 121, 129, 133-135
insulators 14, 15, 30–33
irreversible changes 4, 120, 121

Joule effect/heating 30, 43, 56, 57

kinetic energy 118

laboratory simulation 76, 130, 131
Laplace equations 20, 70

Lennard-Jones potential 89, 90, 101–104, 119, 121
Lifshitz scale/length 26, 27, 43, 44, 127
light-emitting diode 76, 77
linear
  chain 130
  crack 86, 91, 108

mantle (earth's) 128
minimum gap 71
modes of loading 80
molecular dynamics simulation 89–91, 119–121
Monte Carlo method 10, 101
most probable
  defect size 38, 39, 65
  failure strength/field 25, 26, 40, 41, 66
  fracture strength/stress 25, 26
  fuse strength/current 36–41
Mott formula 118

networks
  conducting 1, 2, 14
  dielectric 64–78
  elastic 2, 6, 15
node-link-blob model 12, 13, 15, 17, 19
noise, $1/f$ 28
nonlinearity 69, 75, 119
nucleation of fracture/breakdown 1, 3, 81, 90, 101, 104, 117, 127
numerical simulation 45–48

pair
  correlation function 8–10
  connectedness function 8–10
percolation 5-20, 26, 27, 41–44, 65–73, 95–106
  cluster 7, 8, 11
  continuum 18–20, 42, 43, 67
  correlation length 9, 13, 24
  exponents 8–10, 15, 18
  lattice 7, 36–42, 64–67
  threshold 2, 5, 7, 11, 17, 24, 31, 33
plastic yield 105, 106
power spectrum 28, 149
precursor effect 121–125
propagation velocity 118
pulse 145–148

quenched disorder/impurity 5, 126

random resistor network 14, 37

Index

random spring network 15
red bonds 12, 17
renormalisation group theory 11
reponse function (behaviour) 145–148
reversible changes 41, 120, 121
rigidity (failure) modulus 16, 102
roughening exponent 91–95, 121

sample size, influence 25, 26, 43, 44, 66, 107
sandpile models 27–29, 130, 140, 146–148
scaling
  functions 10, 79
  relations 10, 15, 16
  theory 9
seismology 129
self-affine properties 95
self-affinity 91–95
self-organisation 4, 27–29, 126, 127
self-organised criticality 5, 27, 28, 126, 127, 130, 140–143, 145–149
self-similarity 11, 13
shear deformation 16
shortest path 67, 68
Sierpinsky carpet 13
simple cubic lattice 7, 17, 37
singly connected bonds 11, 12, 58
singularity 2, 3, 18, 26
site disorder 7, 17
slip events 126–132
solid-solid friction 129
sound velocity 118, 123
spanning cluster 7, 11, 12
spring-block model 130–133
spring network 17, 91, 103
square lattice/networks 7, 26, 76

statistics
  cluster 7, 8
  extreme 22–27, 106, 107
  percolation 5, 7, 8, 26–27
stick-slip model 129, 130–133
stress 1, 81, 82, 115–117, 95–109
stress concentration 5, 20–22, 24, 79
stress-strain relation 1, 81, 82, 115–117
superconductor-conductor network 14, 16
super-elastic network 16
superlattice 13
Swiss-cheese model 18, 19, 42, 104, 105

tectonic plate 127, 136
  motion 127
  velocity 127, 136
tip-splitting 120
travelling wave solution 135, 138
triangular lattice/network 7, 91, 101–104

ultrasonic emission 125–126
universal value/constant 121, 124
universality 6, 9, 17

Verlet scheme 90

Weibull modulus/parameter 24, 107, 108
Weibull statistics/distribution 3, 24–26, 40, 107–113, 117

yield stress 100, 101, 105, 106

Zener breakdown 1, 2

The manufacturer's authorised representative in the EU for product safety is
Oxford University Press España S.A. of el Parque Empresarial San Fernando de
Henares, Avenida de Castilla, 2 – 28830 Madrid (www.oup.es/en or product.
safety@oup.com). OUP España S.A. also acts as importer into Spain of products
made by the manufacturer.

www.ingramcontent.com/pod-product-compliance
Lightning Source LLC
LaVergne TN
LVHW011001250326
834688LV00003B/57